Military Publications

Richard Weiner

Other Books By Richard Weiner:

News Bureaus in the U.S.
Professional's Guide to Public Relations Services
Professional's Guide to Publicity
Syndicated Columnists

Library of Congress Catalog Card Number: 78-64584
ISBN 0-913046-09-4

Printed in the United States of America
Published by
Richard Weiner, Inc.
888 Seventh Ave., New York, N.Y. 10019

"Stars and Stripes, G.I. Paper, in Financial Trouble."

This headline appeared on June 20, 1978, in The New York Times and other newspapers. The Associated Press article stated that The Stars and Stripes, the American servicemen's daily newspaper abroad, is in financial trouble and may have to carry commercial advertising to avoid possible extinction.

To millions of veterans of four wars, the article evoked a variety of memories—of cartoons, photos and articles which appeared in The Stars and Stripes and in other military publications.

Just about every military office, base, ship, agency and unit of any kind has its own house organ or other publication, ranging from mimeographed, single-sheet newsletters to professionally produced full-size newspapers and magazines. Military publications, many of which accept advertising, render an important service to their readers, and their cumulative audience (and cost of publication) still is sizable.

It therefore is surprising that there has not been a survey book or directory of military media.

U.S. military forces currently are slightly over two million men and women. Though this number is considerably less than in wartime, it's a sizable group. A very large percent of peacetime military personnel are married and have children. In addition, many civilians are employed at military installations or work for military contractors. Thus, our current military population is several million people.

Military personnel and their families generally read the same magazines, listen to radio and watch the same television programs as other Americans. In addition, over 1000 newspapers and magazines are published exclusively or mostly for military personnel. These publications are extremely important in communicating with military personnel and their families. Many advertisers, publicists, free-lance writers, journalists and other communicators are unfamiliar with military media. The goal of this book is to fill this void.

The 1977 Army population of 761,623 consisted of 84,984 officers (79,599 men and 5,385 women) and 676,639 enlisted personnel (631,410 men and

45,229 women). Plus their families.

The 1977 Navy population of 520,992 consisted of 64,384 officers and 456,608 enlisted personnel.

The 1977 Air Force population was 580,956 and the Marine Corps totaled 192,000.

In 1945, the U.S. Army population (including the Air Force) was 8,266,373 and the Navy was 3,380,817. The number of military publications was well over 2,000, ranging from one-page mimeographed newspapers, to super-professional operations such as The Stars and Stripes and Yank, each of which had circulations in excess of two million.

Local military publications, particularly the base newspapers, are extremely important because many military personnel do not read the local daily newspapers, as they generally are not interested in community news. About 14% of military personnel are officers, and this group has a higher proportion of readers of local daily newspapers.

The lead item in the June 5, 1978, issue of The Gallagher Report, a marketing newsletter, was headlined, "Competition To Heat Up In $7.5 Billion 'Military Consumer' Market," and stated:

GR predicts advertiser battle for position in military marketplace to intensify as military "retailers" (commissaries, post exchanges) grow in marketing sophistication. At stake: estimated 11 million military consumers (over 50% military retirees). Prime focus on 418 commissaries throughout the world (military "supermarkets" generate estimated $3.3 billion sales). Remaining $4.2 billion sales from post exchanges (military "department stores" specializing in hard goods, home furnishings).

The newsletter then discussed the military sales efforts of Proctor & Gamble, Lever Bros. and other companies. At Colgate-Palmolive, The Gallagher Report estimated that $5 million currently is spent annually in military market advertising, mostly in base newspapers and such magazines as Stateside Family, Overseas Family, Armed Forces Network TV Guide and Pro Mag.

The U.S. Army currently publishes about 250 newspapers, including civilian enterprise publications. In addition, about 100 Army Reserve and about 30 National Guard publications are published.

Enrollment in the Reserve Officers Training Corps (R.O.T.C.) recently has increased, from a 1973 low of about 50,000 to about 85,000 college students in 1978. About 60,000 are in the Army program, about 8,000 in the Navy and about 17,000 in the Air Force. About 20% are women.

Most of the military publications listed in this directory are weekly newspapers published for a specific military base. Three of the newspapers are dailies, including The Stars and Stripes, which now has a paid circulation of about 145,000.

The largest-circulation publications are the magazines of veterans and reserve organizations, particularly the national and state magazines of the American Legion, and the national and women's auxiliary magazines of the

Veterans of Foreign Wars.

Other circulation leaders are Army Reserve Magazine (circulation 585,000, an official publication published at The Pentagon), Ladycom (circulation 400,000, privately owned, for wives of military personnel) and two magazines for retired personnel (Retired Naval Personnel and Retired Officer.)

A few of the magazines are bimonthly or quarterly and a few are annuals, including the Tyndall Guide (circulation 15,000, published in Panama City, Florida) and the Strategic Air Command Base Guide (an influential circulation of 8,000, published in Omaha, Nebraska). Several newspapers have circulations of 1,000 or less. One of the smallest is The High Flyer, a 20-page monthly at Fort Bliss, Texas, with a circulation of 150.

Only a few of the publications subscribe to the Audit Bureau of Circulation or other independent auditors of circulation, and circulation figures generally are those provided by the publisher. In a few cases, these may be inflated. On the other hand, the figures sometimes are on the low side, such as in the case of naval and other official newspapers where the circulation varies at different times of the year. The average circulation of the civilian enterprise newspapers is about 10,000.

A few publishers are secretive about circulation and distribution data, for military or other reasons. For example, the rate card of the Defense Transportation Journal includes detailed information about advertising sizes and costs but omits the circulation (estimated at 11,500).

In many cases, the print order is determined by the number of personnel at the military facility, with one copy generally printed for about three or four individuals. For example, about 1,700 copies of the Langley Outlook are printed at Langley Air Force Base in Virginia—one copy for every three employees. About 4,000 copies of Raider Magazine are distributed biweekly to the American military community of about 15,000 people at Torrejon Air Base in Madrid, Spain.

The 1:4 ratio is fairly common. Tyndall Air Force Base in Florida prints 350 copies of Civilian Courier, an eight-page monthly, for about 1,300 civilians at the base. Interceptor Magazine at Peterson Air Force Base in Colorado prints 4,450 copies each month, but the publisher estimates a readership of 6,000.

SAC Personnel Brief, published bimonthly (circulation 500) by the Strategic Air Command at SAC Headquarters, Offutt Air Force Base in Nebraska, prints its distribution list on the inside back cover of each issue.

Ketch 22 is the delightful name of a quarterly newspaper published by the Navy in Seattle, Washington. About 2,000 copies are printed for 4,000 Naval Reservists. The Marine Corps in Washington, D.C., prints 83,000 copies of its quarterly Newsletter for Retired Marines.

Most of the Government publications are restricted in their circulation to military personnel who are part of the unit or relevant to the subject. Occasionally, a publication, or a specific issue, is classified confidential. For example, some issues of the USAF Interceptor Weapons Newsletter are classified confidential and distributed carefully. (Circulation of this quarterly, published at Tyndall Air Force Base in Florida, is 350.)

An example of a large-circulation publication which is "tightly distributed" is the U.S. Air Force Policy Letter for Commanders published semimonthly at Air Force Headquarters in Washington, D.C. Circulation of this four-page newsletter is 82,000. The Air Force also publishes a monthly supplement to the Policy Letter for Commanders in a more detailed, 36-page version, with a circulation of 31,000.

However, many military publications are available to veterans, suppliers and the general public. In fact, subscriptions often can be obtained from the Superintendent of Documents in Washington, D.C. And, of course, subscriptions to privately printed publications, such as Army Times, are available directly from their publishers.

Librarians, particularly those in areas with concentrations of military families, often subscribe to military publications. College librarians subscribe to engineering and other technical military publications, such as the Navigator Magazine published at Mather Air Force Base in California.

The Naval Resale System office in Brooklyn, N.Y., prints 6,000 copies of its News Digest Magazine for about 31,000 employees and vendors throughout the world.

One of the most unusual of the special-interest publications is European Scientific News, published monthly in London by the Office of Naval Research.

The U.S. Army Center of Military History in Washington, D.C., publishes the Army Museum Newsletter for the people in the Army Museum System and others responsible for the safekeeping of Army historical items. About 1,200 copies are printed two to four times a year, and editor J.H. Ewing provides a limited number of copies to educators, museum directors, librarians and military historians.

A few government publications are specifically published for civilian readers. Water Spectrum, a 14,000-circulation quarterly published by the Army Engineers, does not cover military matters and in essence is the Corps of Engineers civil works journal of water resources management.

Many of the Government publications are widely read and highly respected among educators and practitioners in related fields. Examples are The Air Force Law Review, published quarterly by the Air Force Judge Advocate General School at Maxwell Air Force Base in Alabama, and U.S. Navy Medicine, published monthly by the Department of the Navy in Washington, D.C.

The letterhead of the Headquarters Strategic Air Command in Nebraska has printed at the bottom, "Peace . . . is our profession."

Sometimes it seems that one of the major activities of the Department of Defense is the publishing profession. No one can fault the DoD for not encouraging comprehensive communication. Many military units have a publication for commanders, another for "the troops," a different one for families and one for civilians.

Among the several publications of the U.S. Military Academy at West Point is an unusual newsletter published five times a year by the Admissions Office. About 2,000 of the 2,300 copies are for members of the admissions field force. A recent issue featured an article about English standards for admissions, advice about finding desirable candidates for admission, news about West Point and admissions data. (Generally, about 12,000 files are started, about 6,000 actually apply and about 1,000 are offered admission.)

Among the military publications, production, design and editorial content vary considerably, and it is unfair to compare highly professional, big-circulation magazines to typewritten, mimeographed newsletters. Let's take a look at one of the outstanding publications—Soldiers, the official U.S. Army magazine published under supervision of the Army Chief of Public Affairs in Alexandria, Virginia.

As stated on its contents page, Soldiers provides "timely, factual information on policies, plans, operations and technical developments of the Department of the Army and other information on topics of interest to the Active Army, Army National Guard, Army Reserve and Department of the Army civilian employees. It also conveys views of the Secretary of the Army and the Chief of Staff on topics of professional interest to Army members and assists in achieving information objectives of the Army."

A recent winter issue featured articles about Space Available air travel (still free within the U.S., $10 for overseas flights), credit, a six-page portfolio by a talented Army photographer, well-written, magnificently illustrated features about skiing and travel, as well as news sections and various regular departments. An article by staff member SP5 Manuel Gomez about the Panama Canal ("The Big Ditch Revisited"), which could have appeared in a major general-interest magazine, started enticingly with, "The Spanish thought about it; the French tried it; the United States did it."

Congratulations to Editor-In-Chief Lt. Col. James E. Witek and his staff, particularly art editor Tony Zidek!

Army buffs can subscribe to Soldiers by sending $18.80 annually to the Superintendent of Documents, U.S. Printing Office, Washington, D.C. 20402.

Following are descriptions of other representative military publications.

Started in 1921, Infantry is "the professional magazine for infantrymen." Of its 25,000 circulation, about 18,000 are mailed free to military units and agencies (i.e., Infantry and Infantry-related units in the Active Army and the Reserve Components), and about 7,000 are mailed to subscribers. The publisher is the U.S. Army Infantry School at Fort Benning, Georgia.

This excellent bimonthly magazine originated as the Infantry School Mailing List. For $1.50 a year, subscribers received Tactical Problems and other pamphlets and instructional materials published by the Army Infantry School. After about 10 years, the school materials were collected and published semiannually in bound volumes. In 1947, the Mailing List evolved into the Infantry School Quarterly; in 1957, the title was shortened and in 1959, the publication frequency was increased from four to six times a year.

A major publisher of official publications is the Naval Safety Center at the Naval Air Station in Norfolk, Virginia. In cooperation with the U.S. Air Force, the Naval Safety Center publishes Navy Lifeline (The Navy Safety Journal), Approach (The Naval Aviation Safety Review) and Mech (The Naval Aviation Maintenance Safety Review). The Center also publishes Fathom (The Surface Ship and Submarine Safety Review). Mech and Fathom are quarterlies, and Navy Lifeline and Approach are monthlies.

Navy Lifeline, which has the largest circulation (about 40,000), is mailed free to provide safety and occupational health information to commanding, safety and medical officers, and supervisors. The magazine is published by the Navy in cooperation with the Air Force and maintains an office at the Air Force Inspection and Safety Center at Norton Air Force Base in California.

As with many official publications, civilians may subscribe ($5.25 a year, to the Superintendent of Documents, Government Printing Office, Washington, D.C. 20402). Navy Lifeline can be extremely useful to factory managers and others in industry. A recent issue included helpful articles about trench digging, ship-to-shore power safety, frostbite, laser safety and jammed weapons.

A few publications are trade magazines devoted to the military market. For example, Defense Transportation Journal is published by the National Defense Transportation Association for its members who are concerned with military traffic distribution and logistics.

The Navy Resale System Office is extremely important to businesses which sell, or would like to sell, to ships and commissary stores and Navy exchanges. The Office, located at Third Avenue and 29th Street, Brooklyn, New York 11232, functions as a "corporate headquarters" to provide many thousands of retail products and services for Navy personnel and their families. Phyllis Starker, who formerly was a medical editor and a college public relations manager, is editor of the monthly NRS News Digest. A booklet, "A Guide for Doing Business," can be obtained from the NRS office.

Though unofficial publications are not published by the U.S. government, they sometimes have close links with military organizations and thus are confused with government publications. For example, take a look at the Marine Corps Gazette and Leatherneck Magazine. Both are published by the Marine Corps Association (which includes the former Leatherneck Association). Headquartered in the old "post headquarters" at Quantico, Virginia, the MCA is a professional association for current and former Marines. The annual dues ($7 for enlisted Marines and $10 for officers) include a subscription to the Gazette (the MCA "official journal") or Leatherneck (for enlisted Marines). Members can receive both magazines for $5 additional. Formed in 1913, the MCA fosters the spirit and preserves the traditions of the Marine Corps. MCA also sells travel, insurance and other products and services to its members.

Other major association publications include Army, published by the Association of the U.S. Army, The Retired Officer Magazine, published by The Retired Officers Association, and The National Guardsman, published by the National Guard Association of the United States.

Army has a monthly circulation of 85,000 to Army officers, members of the Army National Guard and Reserve and civilians with a strong interest in national defense.

The Retired Officer has a circulation of 265,000 to officers of the *seven* uniformed services (Army, Navy, Air Force, Marine Corps, Coast Guard, National Oceanic and Atmospheric Administration and Public Health Service). The $10 annual membership in The Retired Officers Association includes a subscription to the monthly magazine and group buying privileges for automobile leasing, rental, insurance, travel and other products and services.

The Retired Officer is a well-produced slick monthly magazine which reports on legislation and other news about military retirement, officers' rights and benefits, and veterans' affairs. Since most of the readers are either in or planning a second career, coverage of business and career opportunities is featured, in addition to articles which relate to their cultural interests, including history, travel, hobbies and book reviews. As might be expected, the largest circulations are in California (46,387), Virginia (21,361) and Texas (19,570).

The National Guardsman, the publication for the Army National Guard and Air National Guard, has a circulation of 65,000. Current membership of The National Guard is about 400,000, of which about 25% is the Air Guard. Women make up about 4% of the membership.

The Marine Corps Association has two monthly magazines, Leatherneck ($1 a copy, $7 a year) and Marine Corps Gazette ($1 a copy, $10 a year). The Gazette is more oriented to officers and has more news and serious articles than Leatherneck, which has a larger circulation and includes cartoons and pinup photos. Both publications were started in 1917.

Among the academic publications, the leading nongovernmental publisher operation is located at Kansas State University in Manhattan, Kansas. The Department of History, which coincidentally is located in Eisenhower Hall, is the home of the American Military Institute, which has published the quarterly Military Affairs since 1937, and the Air Force Historical Foundation, which has published the quarterly Aerospace Historian since 1954. The editor of both journals is Robin Higham, Ph.D.

Since 1968, when Kansas State University took over responsibility for Military Affairs, the journal has been expanded into many different fields of military history, plus news of academic courses, museum exhibits, bibliographies, and other subjects of interest to historians, political scientists and other civilians and military personnel. Membership in the American Military Institute ($7.50 a year for students, $15 for other individuals and $20 for institutions) includes a subscription to Military Affairs. About half of the subscribers are libraries and other institutions.

A recent issue of Military Affairs included book reviews, academic news, a bibliography and the following articles:

- On The Infantryman's Age In Eighteenth Century Prussia
- British Intervention In Russia During The First World War
- Saint Barbara's Legion: Portuguese Artillery In The Struggle For Morocco, 1415-1578
- Antecedent Education And Officer Recruitment: The Origins And Early Development Of The Public School-Army Relationship
- 1861: Blockade Vs. Closing The Confederate Ports
- Abilene, Kansas, And The History Of World War II: Resources And Research Opportunities At The Dwight D. Eisenhower Library
- The AMI Constitution Amendment: General Purposes For The Recommended Changes To The By-Laws Of The American Military Institute.

Most of the authors of these articles are history professors. In this issue, for example, the authors are a University of New Orleans specialist in the 18th-century Prussian army, a Marshall University specialist in the First World War, a University of Virginia specialist in Brazilian and early Portuguese history and an archivist at the Dwight D. Eisenhower Library in Abilene, Kansas.

Aerospace Historian, published since 1970 at Kansas State University, is the international journal of aerospace history, with occasional articles on military and naval history and theory. Unlike Military Affairs, the majority of subscribers are individuals (rather than libraries or college departments), most of whom are current or retired Air Force officers. Members of the Air Force Historical Foundation ($15 a year) receive its official journal, Aerospace Historian, a quarterly magazine which is livelier in artwork, writing style and subject matter than the more scholarly Military Affairs.

Subjects span the spectrum of flight, aerial warfare and aircraft from the Civil War to the aerospace program. Authors include pilots, wartime leaders, explorers, asronauts, scholars and others who manifest our conquest of the sky.

The lead article in a recent issue of Aerospace Historian, by Lt. Col. John H. Napier III, command historian at the Air University, was about the symbiotic relationship between Montgomery, Alabama, and Maxwell Air Force Base (which includes the Air University, and its publication, Air University Review).

Other articles included histories of the Turkish Air Force, the 1st Aero Squad (formed in 1913, the oldest air arm of the Air Force), the Air Rescue Service, photos of historic British planes and an ode to "Jenny" (the Curtiss JN two-seater biplane trainer.)

About 17% of the U.S. military population is black. The largest numbers of blacks—195,027, or 25.1% of 777,000—are in the Army. The second largest concentration—77,000, or 22% of 350,000—is in the National Guard. The smallest percentages of blacks are in the Navy (39,480, or 7.5% of 524,875) and Coast Guard (2,240, or 7.2% of 31,287). To complete these late 1977 figures, the number of black Marines is 30,720 (16% of 192,000) and black Air Force personnel total 72,877 (12.7% of 575,525).

In 1977, the total number of black military personnel, including the National Guard, was 417,344. This is a sizable audience, and it was inevitable that a publication be launched to reach this group. In early 1978, Tanner Publications, publisher of Encore American (a biweekly black news magazine with a circulation of 150,000), introduced Eagle & Swan, a quarterly magazine for black men and women in the U.S. Armed Forces. Single-issue price is 75 cents.

The eagle is a traditional symbol of strength, and, as noted on the cover, the swan is "a tenacious fighter, sacred to the Ancients." The slick magazine reports on accomplishments of black service people, features news about reenlistment bonuses and other career information, and also has departments on sports, books, fashion and the arts.

* * *

Several of the official military newspapers do not accept news releases or other publicity material from any source outside of the Department of Defense. However, many official publications welcome by-lined articles, letters and other contributions from readers. For example, Navy Civil Engineer, a quarterly magazine published by the Naval Facilities Engineering Command (and printed at the Naval School, Port Hueneme, California) consists almost entirely of by-lined articles by current and former naval engineers.

Regardless of whether a local publication is government or privately owned, the news content generally is compiled by military personnel who are located at the public affairs office of the military installation.

News input is provided by many sources, including Commander's Digest (published by the Department of Defense), Armed Forces Press Service (AFPS), Army News Features, Army News Photo Features, NAVNEWS, Air Force News Service (AFNS), International Service Note (Marine Corps), clip sheets and other materials from the North American Defense Command, Strategic Air Command, Air Force Logistics Command and other major commands, as well as the Civil Service Commission and government groups.

The AFPS in Arlington, Virginia, publishes 2,500 copies of its weekly Clipsheet for military editors in all branches. The Department of the Army has a bimonthly newsletter for public affairs officers.

The Defense Information School at Fort Benjamin Harrison (near Indianapolis) has published since 1974 the quarterly Military Media Review. Current editor is Connie McKean.

The Armed Forces Information Service of the Department of Defense publishes many useful aids for editors of military publications, including the

Armed Forces News Style Guide. This 54-page booklet follows the style of The Associated Press and United Press International, and also includes an 18-page section about military ranks, titles and abbreviations.

Service newspaper editors are instructed to devote particular attention to the I-O-U Department, that is, the Ins, Outs and Ups of military personnel.

According to the Applied Journalism Handbook (published by the Defense Information School at Fort Benjamin Harrison, Indiana 46216), "the military journalist is a combination of skilled journalist and military person. Those who work on service newspapers are akin to local or community newspaper reporters, with two basic missions:

"One of the missions is to enhance the morale of the organization it serves. High morale results from many factors, not the least of which are knowledge and esprit de corps. Newspapers can enhance the morale of any organization by keeping their readers informed about off-duty recreational activities and achievements of units and individuals, to name just two.

"The other mission of the service newspaper is to serve as a channel for the commander to put forth his policies to his command.

"It is through the newspaper that the commander informs his personnel of safety programs, voting activities and educational activities. It is through the newspaper that the commander explains the effects of new personnel programs and training activities. It is through the service newspaper that the commander implements, at least in part, his internal information program."

As part of its program to encourage high-quality, professional journalism, the Army Chief of Information presents an Army Journalist of the Year Award, which is akin to a Pulitzer Prize for Army Journalists. Started in 1975, the award is named in honor of Paul D. Savanuck, a Stars & Stripes journalist who was killed in Vietnam. The first recipient of the award was John M. Wells, a reporter for Medcom Examiner, a 12-page, monthly newspaper published in Heidelberg for the Army Medical Command in Europe. Like Staff Sargeant Savanuck, Specialist 4 Wells was a graduate of the Defense Information School.

Medical publications have a special appeal. The 1976 Savanuck Journalist was Sp 5 Nonda J. Duerschmidt, a staff writer for Stripe, the newspaper of Walter Reed Army Medical Center in Washington, D.C. Miss Duerschmidt, 24, who was in the Army for six years, won the award for articles about cancer, marijuana, homosexuality and other controversial subjects.

The 1977 winner was Spec. 5 Gary L. Bloomfield, editor of Indianhead, the newspaper of the 2nd Infantry Division in Korea.

The Army also recognizes excellence in journalism with the Keith L. Ware awards, established in 1970 in memory of a chief of information who was killed in combat in Vietnam. Winners in various categories have included Castle Courier, Guardian, Uptight, Jagged Sword, Illuminator, Third Word, Tour 365, Supthai Sentinel and Bayonet—all of which are no longer published.

The 1977 Ware winners were Maine Guardsman (published monthly by the

121st Public Information Detachment at Camp Keys, Augusta, Maine), Buckeye Guard (published bimonthly by the 196th Public Information Detachment of the Ohio Adjutant General's Department at Worthington, Ohio), Inside the Turret (an unofficial civilian enterprise weekly published at Elizabethtown, Ky., (near Fort Knox), The Aircraftsman (published biweekly at the Army Depot, Corpus Christi, Texas), Persuader (published monthly by the 4th Psychological Operations Group at Fort Bragg, N.C.), Support Sentinel (published biweekly by the 21st Support Command in Kaisenslautern, Germany), Patriot (published biweekly at Headquarters, 24th Infantry Division and Ft. Stewart/Hunter Army Air Force at Fort Stewart, Georgia) and Tropic Lightning News (published biweekly at the 25th Infantry Division, Schofield Barracks, Hawaii).

In an unusually candid article in Military Media Review (Fall, 1974), Navy Journalist Third Class Howard L. Rosenberg described his experiences in changing the design and editorial style of Sealift, the monthly magazine of the Military Sealift Command. The article, delightfully titled, "Two Years Before the Masthead," noted:

"Throughout the services, the military editor is usually young and low ranking. His job generally has responsibility far exceeding that of his peers. He is told he is the spokesman for the commander and the command. The publication he edits must compliment both, yet still act as an information vehicle. How can that be accomplished?

"Rank is one of the most difficult barriers to overcome. There is a necessary communication between the commander and editor that must be kept open. Although the editor is not invited to staff conferences or consulted concerning major command decisions, he is expected to understand and support them through his publication.

"The answer to this problem is an intermediary who *is* privy to conferences and decisions. This intermediary is the public affairs officer, and the editor must maintain a constant and open communication channel with him in order to objectively judge potential news and feature material.

"To do this, the editor must convince the PAO of his integrity and professional ability. Winning that trust, and keeping it, is a never ending job, one of the many which an editor must perform to get his magazine printed."

Making use of his experience at the Defense Information School's courses, JO3 Rosenberg redesigned the magazine not only with new typefaces, covers and logo, but, more important, with livelier art and writing, including the elimination of as much governmentese as possible. Perhaps the biggest shocker was the ban on use of most acronyms unless they are accompanied by an explanation or spelled out. Abbreviations such as PAO (Public Affairs Office) are permitted if commonly understood.

An unofficial service publication, also called by the military a "civilian enterprise," must print a disclaimer stating that it is not an official publication, and its advertising does not constitute endorsement by the armed forces. The publisher of a civilian enterprise newspaper is chosen by bid and various conditions, including annual renewal of the "license agreement," are set forth by the commander of the local installation.

Many unofficial publications are published under military contract and distributed free at post exchanges and other military locations. For example, the Quantico Sentry (circulation 11,000) is published on Fridays in Dumfries, Virginia, and distributed to military and civilian personnel, and their families, at the Quantico Marine Corps Base in Quantico, Virginia.

Many of the unofficial newspapers are similar to suburban weekly newspapers. Some of these editors are interested in receiving "hometown news" and occasionally are receptive to mats and offset proofs from publicists in order to reduce production costs for photos and artwork.

In Virginia Beach, Virginia, for example, the Armed Services Newspaper Corporation publishes four monthly tabloids which are mailed to the many military installations in the seven Hampton Roads cities. Distribution includes the Norfolk Naval Shipyard and Langley Air Force Base. The company, which has been locally owned and operated for over 50 years, utilizes publicity material and also publishes a variety of syndicated columns, mostly from the McNaught Syndicate, including Joey Sasso (Inside TV), Doris Lilly (Gossiping) and Bob Nesoff (Camping).

Several local publications are distributed free to military personnel but are totally civilian in ownership and operation. The Fort Devens Dispatch, for example, is crammed with military news at Fort Devens and also news from Ayer, Fitchburg, Groton, Harvard, Littleton, Pepperell, Shirley, Townsend and other towns in eastern Massachusetts. As a result, the publication looks like a typical small-town weekly newspaper, complete with PTA, women's club and other local news.

The distinction between official and unofficial is further blurred by the newsletters, newspapers and magazines which are "authorized," unofficial publications. For example, The Write-On is an authorized, unofficial Army publication published by the Army Intelligence and Security Command Support Group at Fort George G. Meade, Maryland. An informal, chatty, eight-page newsletter, The Write-On is a labor of love by SP4 Patricia M. Stewart (editor and typist) and SP4 Regina A. Grippin. An alumnus of Fort Devens, Miss Stewart served in Thailand and Korea. If and when she is reassigned, The Write-On may or may not be continued. (It was not published for several months after the previous editor left Fort Meade.)

On the other extreme is The Spangled Banner, a 2,000-circulation, six-page tabloid newspaper which is the official Class II Air Force newspaper published weekly (Fridays) for the personnel of Spangdahlem AB in West Germany. The typesetting (justified on both margins, four columns, extensive use of subheads), editorial and photo quality is first-rate and reflects the staff size. (Though not expensive, it's quite a different operation from The Write-On.)

Staff size is not necessarily a barometer of quality. Army Reserve Magazine is an excellent 36-page, bimonthly publication, with a staff of two. The three-column format and content resembles Time magazine (before its redesign). Though published at the Pentagon, Army Reserve is a low-budget operation. In 1977, one of the issues was dated May, June, July, August and included this explanation:

"MISSING ISSUES: April in Paris, March in early August or other combinations of dates do not explain why this magazine has seemingly lost track of time. This is a combined May-June-July-August only because a researcher-writer-editor-art-director-bureaucrat staff of one cannot go it alone without falling behind. Persistent personnel shortages have kept the long forgotten past before us or put our behinds in front of us, so to speak. Please be patient. It's not the Post Office; it's unspeakable."

Similarly, circulation is not necessarily related to quality. The Relay is a professionally produced, six-page tabloid distributed monthly to the less than 1,000 members of the Army Communications Command in Taiwan. The design and writing style of this authorized, unofficial publication is excellent.

* * *

The largest private publisher in the military field is the Army Times Publishing Company in Washington, D.C. The company publishes Army Times, Air Force Times and Navy Times, weekly tabloid newspapers which have a combined circulation (verified bythe Audit Bureau of Circulations) of 373,982. The company also publishes The Times Magazines, a four-color, 8½ x 11 supplement inserted every other week in the three military newspapers, and special supplements (retirement, education, travel, auto, shopping guides) issued about once a month. In addition, the company publishes Military Market Magazine (a monthly, in two editions—commissary and exchange) and Federal Times (a weekly with a format similar to the military newspapers).

The circulation of the military newspapers—Army Times (142,331), Air Force Times (134,443) and Navy Times (97,208)—is mostly (88 percent) in the U.S.

One indication of the power of these publications is that their circulation is paid, not free, and it's mostly subscription (326,349, at $21 a year!). Average sales of single copies (at 75 cents) are 47,633.

The distribution is to wherever there are Army, Air Force, Navy, Marine Corps and Coast Guard personnel. The largest numbers are in California (first), Texas, Virginia, Florida, North Carolina, Georgia and Maryland. About 22% of the circulation is to retired military personnel.

Outside of the U.S., the largest numbers are in West Germany (20,539, or over half), Korea, United Kingdom, Japan, The Philippines and Italy.

The Times newspapers are a post-World War II phenomenon. The two oldest, Army Times and Air Force Times, were started in 1939. Navy Times followed in 1950. Each has an independent editorial staff.

The president of this extraordinary company is William F. Donnelly. The publisher, Mel Ryder, heads an extremely successful advertising staff. Major advertisers include cigarette companies (Vantage, Salem, Viceroy, Decade, Benson & Hedges, Fact, Kool, Kent, Belair), liquor (Johnnie Walker, J&B, Gordon's, Canadian Club, Old Grand-Dad), photography (Polaroid), automobiles (Triumph), insurance (companies which specialize in military families) and many colleges and educational institutions. Many pages of small-space and classified ads promote schools, travel, career opportunities and real estate. An interesting category of advertiser is the military. For example, full-page ads are placed by Army Reserve.

Of course, the Times newspapers are not all advertising. A typical 64-page issue contains more news than in many daily newspapers, and most of it is staff-written *news,* rather than features.

With regard to features, one of the leaders is "off duty," a monthly magazine which, as implied by its name, is about travel, sports, hobbies and other leisure interests of military personnel and their families. It is circulated free in U.S. Government housing areas and recreational and resale facilities (e.g., post exchanges) on military installations.

With a total circulation of 390,000, "off duty" offers advertisers extraordinary flexibility (though not at cheap rates) by producing three separate editions and then splitting each edition demographically and geographically.

The U.S. edition, circulation 190,000, is distributed in the Western states, Alaska, Hawaii and the Caribbean. There is no distribution east of Montana, but in California advertisers can zero in on, for example, the Camp Pendleton Marine Corps Base (circulation 12,000, $485/page).

The European edition, circulation 120,000, is distributed in Germany, United Kingdom, Benelux, Mediterranean and the Near East, and the Pacific edition, circulation 80,000, is distributed primarily in Japan, Okinawa, Guam, Korea and the Philippines. Among the choices offered to advertisers are sub-editions primarily to women (distributed at family service centers), as differentiated from the sub-editions to troops.

To produce these variations, publisher Walter B. Rios has set up an international operation headquartered in Hong Kong. Editorial offices are in Hong Kong, Frankfurt and Newport Beach, California, advertising offices are in Hong Kong, Frankfurt, Tokyo, New York, Honolulu and Newport Beach, and printing facilities are in Hong Kong.

One of the largest publishers of local military publications is the Southern Publishing Company, 2101 W. Government, Pensacola, Florida. Owned by the Foote family (Eugene Foote is president and Doug Foote is in charge of sales and advertising), Southern produces seven publications for the Naval and Air Force personnel and their families in the Pensacola area.

The largest is the Gosport, a weekly tabloid newspaper which has 14,500 copies distributed free each Friday to the Naval Air Basic Training complex of bases. Most are at the Naval Air Station in Pensacola and the Naval Education and Training Program Development Center at Ellyson Field. These military and civilian personnel have a combined payroll of over $21 million a month!

Founded in 1942, the Gosport takes its name from the voice tube used in the early days of naval aviation by flight instructors to give directions and instructions to students. The voice tube was invented in Gosport (originally God's Port), England.

Here's a summary of the other Southern publications.

The Whiting Tower, the weekly newspaper for the Naval Air Station at Whiting Field in Milton, has a weekly circulation of 4,500. NAS Whiting Field is the home of Training Air Wing Five, where all fixed wing and rotary naval aviators complete their training and receive their wings.

The Eglin Eagle is distributed free as the base newspaper for Eglin Air Force Base, which is the largest Air Force Base in the world. The main Eglin complex is north of Fort Walton Beach and is the largest employer in the "Playground area." Circulation of this weekly tabloid is over 8,000, of which about 2,000 copies are home-delivered to base housing.

SOF Hi-Lites is the base newspaper for Hurlburt Field, which is located in Mary Esther, eight miles west of Fort Walton Beach. This Tactical Air Command Air Force Base, which is the home of the Special Air Warfare Center, trains all Commandos and special units of all military branches. (SOF refers to Special Operations Forces.) The free weekly circulation is 4,800.

Lemot is the weekly publication of the wives of Air Force Officers at Eglin Air Force Base and Hurlburt Field. It reaches active duty and retired officers' wives and has a circulation of 4,500.

The Mustin Missile is mailed monthly to the 5,000 members of the Mustin Beach Officers' Club. Dues are $48/year.

The Officers' Wives' League Bulletin is the monthly publication of the wives of Naval Marine and Coast Guard Officers at the Naval Air Training Command complex of bases in the Pensacola area.

The Southern Publishing Company operation is extremely efficient. The weekly base newspapers are printed offset on 10 x 16-inch, five-column, tabloid-size, high-quality paper and distributed on Fridays, except for Lemot, which is distributed on Mondays. The Mustin Missile is distributed on the first day of each month and the Officers' Wives' League Bulletin is distributed on the 15th of each month. The Mustin Missile, Officers' Wives' League Bulletin and Lemot are 8½ x 11-inch magazines.

The company also does commercial printing, including high shool publications and engraved wedding invitations.

The Pensacola area is in northwest Florida, near the Mississippi cities of Biloxi and Gulfport.

Many official publications do not accept advertising. Those which accept advertising contract with a commercial publisher, generally a nearby weekly newspaper. For example, Mach Meter is the official newspaper of the Headquarters 27th Tactical Fighter Wing at Cannon Air Force Base in New Mexico. The business and production, including advertising sales, are handled exclusively by the Portales News-Tribune Company.

Advertising rates generally are for a single insertion. Some publications provide frequency and/or volume discounts, commissions to advertising agencies, and, occasionally, a lower rate schedule for local advertisers. For example, the Quantico Sentry in Virginia charges national advertisers 28 cents per agate line, or $3.92 per column inch, whereas local advertisers pay $1.80 to $2.90, depending on number of inches per month. Agencies receive a 15 percent commission, plus two percent cash discount if paid by the 10th of the month. Special charges, such as production, preprints, special position, and color, generally are the same for national and local advertisers and are not commissionable. At the Quantico Sentry, the charge for a color in addition to black is $90 for one color, $150 for two colors and $180 for three colors.

* * *

The focus of this directory is on publications for military personnel. Publications for reserve organizations and veterans also are included.

The two largest-circulation veterans publications rank in circulation and advertising linage with Esquire, Playboy and the other top-circulation men's magazines. Specifically, American Legion Magazine has a circulation of over 2½ million, and V.F.W. Magazine has a circulation of over 1½ million. These magazines really are combinations of fraternal and general-interest publications. Military subjects are included, but the emphasis is on sports, economic and social issues of interest to older men, as well as news relating to the organization and its membership.

Veterans organizations range from such elite groups as the Flying Tigers (250 members) and the American Fighter Aces Association (515 members) to the American Legion, with a staff of 281 and over 16,000 local groups. Among the veterans organizations with publications are the Order of Daedalians (pre-World War I pilots and their descendants), Military Order of the Purple Heart of the U.S.A. (12,000 members, staff of four), Catholic War Veterans of the U.S.A. (50,000 members), Children of the Confederacy (10,000 members), Disabled American Veterans (545,000 members, staff of 795), Paralyzed Veterans of America (11,000 members), Jewish War Veterans of the U.S.A. (105,000 members), AMVETS (American Veterans of World War II, Korea and Vietnam, 150,000 members).

Wherever there is a concentration of military personnel the nearby general circulation daily newspapers obviously devote sizable attention to military news. Advertisers, publicists and others interested in the military market should keep this in mind when preparing media lists, though these publications generally are not listed in this directory.

Norfolk, Virginia, for example, is one of the largest markets in the Southeast because of the U.S. Naval Station and other major military facilities. Retail sales of military families (over 200,000 people) in the Norfolk area total about $250 million. The general circulation dailies, The Virginian-

Pilot (circulation 128,000) and The Ledger-Star (circulation 97,000) publish a Sunday edition with a whopping circulation of 194,000. The local military newspapers are Soundings, Langley Flyer and Fort Eustis Wheel, all of which are published by one civilian enterprise company.

San Diego, California, is one of America's largest markets because of its Army, Navy, Marine Corps and Coast Guard installations and schools, including its renowned Naval Hospital. The largest segment consists of Navy personnel. About 90,000 Navy men and women—17% of the entire Navy—are stationed at the 170,000-acre naval complex at San Diego.

The city is the headquarters of Copley Newspapers, which publishes the San Diego Tribune (circulation 128,000) and San Diego Union (circulation 189,000). The circulation of the Sunday edition is an astounding 311,000.

In addition, there are five civilian enterprise newspapers, the Dispatch, Dry Dock, Jet Journal, Shore Leaf and North Islander.

The U.S. armed forces in West Germany can find out about radio and television programs of the Armed Forces Network and other broadcast schedules from two publications, which, like TV Guide, also contain leisure-interest articles. Armed Forces Network TV-Guide, published in Frankfurt, has a circulation of 92,000, and R&R Entertainment Digest, published in Heidelberg, has a circulation of 84,000.

The distribution to armed forces and their families of large-circulation consumer magazines is so sizable that many of them offer special military editions or advertisements with circulation restricted to military readers. The Reader's Digest had an overseas military edition. Time, Newsweek and Sports Illustrated offer an advertising package with a combined circulation of about 104,000, as do Family Circle and Woman's Day, with a combined circulation of about 200,000.

Media Networks (600 Third Avenue, New York 10016) is the advertising representative of a group of 11 men's magazines with a combined military circulation of about 140,000 and six women's magazines with a combined military circulation of 99,000.

* * *

The data in this directory was compiled by sending questionnaires to several thousand publications. Most of the publications were on lists purchased from the Department of Defense.

In a conscientious attempt to obtain information from nonrespondents, the questionnaire was mailed a total of three times. In several cases, editors referred the questionnaire to officers in New York and Washington, D.C., who called for more information about the book. A few editors declined to provide data, because of military secrecy, privacy, or other reasons.

Of course, we respect all of these requests and have not listed either publications which did not return questionnaires or publications which requested anonymity.

Occasionally, military publications are started, terminated, merged and changed, though not as often as civilian publications. In late 1977, the Army Intelligence and Security Command replaced The Intelegram with the Journal. The news magazine has more features of interest to the Army Intelligence Community than The Intelegram, which was more of a newsletter. Prior to Intelegram, the Army Security Agency published Hallmark. More important, the new Journal reflects the Army's grouping of strategic intelligence and counterintelligence into a single new command, INSCOM.

Among the military publications which have been terminated in recent years are:

Adcom Surgeon's Bulletin, Peterson Air Force Base, Colorado
Army ROTC Newsletter, Fort Monroe, Virginia
Belvoir Bluejacket, Fort Belvoir, Virginia
Camp Foster Scoop, Pacific Fleet Marine Force
Civilian Personnel Newsletter, Defense Nuclear Agency, Washington, D.C.
Commander's Newsletter, Ellington AFB, Houston, Texas
Employee's Newsletter, Ellington AFB, Houston, Texas
HQ Command USAF Personnel Management Bulletin, Bolling AFB, Washington, D.C.
Korat CBPO Newsletter, APO, San Francisco, California
Management and Automation Digest, Washington, D.C.
MOCAS Memorabilia, Alexandria, Virginia
Naval Weather Service Command Newsletter, Washington, D.C.
Safety Fuse, Defense Electronics Supply Center, Dayton, Ohio
Safety Newsletter, APO, N.Y.
Taiwan Tabloid, APO, San Francisco
Tapo Review, APO, San Francisco
TFW Aircraft Evaluation Newsletter, APO, San Francisco
Thai Employee Newsletter, APO, San Francisco
U.S. Civilian Notes, APO, San Francisco

In 1978, the former editor of DoD Productivity Newsletter (formerly published by the Defense Industrial Resources Support Office at Cameron Station in Alexandria, Virginia), plaintively noted, "This publication currently in suspense."

Perhaps the Department of Defense decided that the Productivity Newsletter was unproductive. However, it is not proper to be facetious about the suspension or termination of a publication. In fact, several recently discontinued publications appear intriguing, particularly Career Motivation Advisory (formerly published by Headquarters Strategic Air Command at Offutt Air Force Base in Nebraska) and High School News Service Report

Magazine (formerly published by the Fleet Home Town News Center at the Great Lakes Naval Station in Illinois). The H.S. Report has returned.

In June 1978, the Defense Department launched a colorful, spritely monthly tabloid, called SAM, to provide "information and entertainment to improve the morale and welfare of U.S. military personnel, their dependents and DoD civilian employees."

Actually, the name should be SSAM, because the acronym refers to Soldier, Sailor, Airman and Marine.

Whereas dozens of base newspapers look alike in their old-fashioned weekly newspaper format, SAM is unique. It's for military personnel who are young and represents a noble experiment to be "with it."

The origin of SAM was stated in the first issue:

"DoD for 12 years has published Commanders Digest for its senior military leaders. That publication—now retitled Command and given a new typographical and editorial look—delivers to the service's field grade officers and top three non-commissioned officer ranks the latest major Defense policy on everything from the All-Volunteer Force to strategic nuclear parity.

"However, the editors of Command have an itch that tells them most young service persons couldn't really care less about topics like strategic nuclear parity. And, a lot of other information young people get in official military publications is often so candy-coated or written in officialese that it's hardly worth reading."

It's too early to tell if SAM will succeed as a pleasant, informative, entertaining fellow.

The articles are short and punchy, the layout relies heavily on color and black-and-white mixtures of drawings and photos, and the publication is sort of a mixture of Mad Magazine, National Lampoon, Rolling Stone and National Enquirer. Among the dozens of serious subjects covered in the first issue were suicide, homosexuality, fraternization (between officers and enlisted personnel) and battered wives. However, the accent is on the lighter side, with articles about Bob Hope and Elvis Presley, comic strips and headlines with lots of puns.

The publication probably will settle down and develop a contemporary style which is enjoyable to young adults without being juvenile. In a boxed item, the editors of the experimental free publication pointed out that funding came from the cancellation of another DoD publication, the Defense Information Guidance Series.

Here's the 1979 distribution of SAM: 65,000 to the Army, 56,000 to the Navy, 72,000 to the Air Force, 40,000 to the Marine Corps and 3,000 miscellaneous, totaling 236,000.

During the Vietnam War, dozens of independent publications were published as an alternate communications network offering antiestablishment views. In 1973, the magazine Military Journalist reported that "the military underground press is alive and well and presents a constructive challenge to authorized military newspapers." Only a few alternative publications have survived the cutback in military forces.

Indicative of changes in numbers of military personnel (and hence in numbers of publications and their circulation) are the following comparisons between 1964 (pre-Vietnam) and 1970.

Total personnel went from 2,691,000 to 3,294,000. The major increase was in the Army, which went from 968,000 to 1,431,000.

For several years after the Second World War, Con Gebbie published a free directory of alumni of Yank, The Stars and Stripes and the Army News Service. A former Yank and Army News Service journalist, Mr. Gebbie became hooked on directories and left his job at Shell Oil Company to start a directory-publishing company, specializing in employee house organs.

Second World War military journalists included cartoonists Bill Mauldin and Herb Block (better known as Herblock); Igor Cassini,a Stars and Stripes reporter who became a jet-setter gossip columnist; critics Howard Taubman, Richard Coe and John Crosby; Fred Friendly; Marion Hargrove (author of "See Here, Private Hargrove"); E.J. Kahn of The New Yorker, and columnists Peter Lisagor and Jimmy Cannon.

Arthur M. Schlesinger (father of historian Arthur M. Schlesinger Jr.) reviewed the military media of the First World War in an article titled "The Khaki Journalist," in the Mississippi Valley Historical Review.

Quite a few military publications were launched during the First World War. In fact, every war has stimulated the birth of military media. The Army and Navy Journal was started in 1863 with the objective of helping "the Union war effort by packing military news in its proper perspective." It's still performing this chore, except that it's now called Armed Forces Journal, which is a lot simpler than several of its interim titles. (In 1962, it was called the Army-Navy-Air Force Journal and Register and was published weekly.)

Though its cover slogan is "The Defense Spokesman Since 1963," the Armed Forces Journal (now a monthly) is privately owned and is frequently critical of the Department of Defense, particularly about budgets. A lengthy article in August 1971 criticized the Pentagon for excessive expenditures on publications. Based on an extensive survey by the Journal, writer Bernadine M. Kopec estimated the annual cost of the Defense Department magazine empire to be about $57 million. She criticized many of the magazines which were published as part of a "publish or perish" academic philosophy.

Defense Secretary Melvin Laird reacted with a review of DoD periodicals. In 1972, the DoD task force reported that some savings could be made, a few of the 1402 periodicals could be combined or eliminated and others could be improved, particularly in level of credibility and comprehensiveness of coverage.

Officers and enlisted personnel who were surveyed reported that civilian publications, notably Army, Navy and Air Force Times, were considerably more useful and well-read than such official publications as Soldiers, All Hands, Airman, Air Force, Air University Review and Command Digest.

One of the publications no longer published but still remembered with great fondness by many veterans of the Second World War is Yank, The Army Weekly, which was produced by and for enlisted men from June 17, 1942, until December 28, 1945. The staff of combat correspondents, photographers and artists put out 23 regional editions from bureaus in Paris, Honolulu, New York and elsewhere which still rank as among the most distinguished and exciting newspapers ever published. Its staff included Pvt. Irwin Shaw, Sgt. Merle Miller, Sgt. Marion Hargrove, Pvt. William Saroyan, Sgt. Dave Breger and hundreds of other G.I. Joes.

Yank was an extraordinary feat in that it was an independent publication started by the War Department but totally unmilitary in its informality. Headquarters was at 205 East 42 Street, New York (across the street from the New York Daily News), where for years the managing editor was Sgt. Joe McCarthy. The magazine sold for five cents and it had a paid circulation of more than two million. George Baker's "Sad Sack" became the most popular cartoon character, almost its symbol.

Similarly, The Stars and Stripes is a phenomenon of the American free press. Established as a weekly in the First World War (February 8, 1918) and revived in the Second World War (April 18, 1942), The Stars and Stripes was described by London co-editor Bob Moora as "a paper for Joe; after that it's a newspaper; after that it's a trade journal whose specialized readers are soldiers, and after that it isn't anything no matter what the brass says."

On November 2, 1942, Stripes became a daily, and it has remained that way. During the Second World War, four- to eight-page editions were printed at 25 locations in Europe, North Africa and The Middle East. Here are excerpts from its "official history."

"To get the wartime job done, staffers conned, cajoled or commandeered printing presses on the war fronts. Stripes was published from friendly newspaper offices, from mobile plants on the battlefield, and from confiscated enemy presses.

"The staff was always on the move. To publish a daily for combat troops, they had to set up shop as close to the front as possible. The best they did was in Rome. They arrived the same day the city was captured, grabbed the offices of Il Messaggero, and the first edition came off the press that afternoon with the banner headline: 'We're in Rome' . . .

" 'Throughout wartime operations,' said Moora, 'the staff had to fight to withstand the pressures of well-meaning but ill-advised officers to use the paper for propaganda purposes, for personal publicity, or as a headquarters directive' . . .

"The first plant on German soil was set up in a Pfungstadt brewery on April 5, 1945. A year later, Stripes moved to Altdorf, taking over the presses of 'Der Sturmer,' a Nazi newspaper.

"Stripes later returned to Pfungstadt. Then on Sept. 27, 1949, it settled in Darmstadt at a former Luftwaffe training field, where it remains. Stripes became a nonappropriated fund agency, which meant it had to pay its own way, and began to civilianize. As wartime military staffers left, Stripes built a staff of full-time professional civilian newsmen, although the paper retains a small military cadre. Today, Americans and local national civilians work together to gather and write the news, print the paper, man the bookstores, and keep the accounts.

"But it's still the service member's newspaper, dedicated to keeping soldiers, sailors, airmen, and Marines serving abroad informed of what is going on in the world."

The European edition of Stars and Stripes is currently published as a 28-page daily tabloid (40 pages on Sunday), with news about U.S. military personnel in 22 countries. Average daily paid circulation is 101,000, at 15 cents a copy. During the Second World War, circulation peaked at slightly more than 500,000. However, it was free.

Under the supervision of Colonel Billy E. Spangler, editor-in-chief of the European Stars and Stripes, the newspaper has retained its editorial vitality. News sources now include its own staff, The Associated Press, United Press International, The New York Times News Service, Editors' Press Service (a syndicate which represents King Features, Chicago Tribune-New York News, Los Angeles Times and other syndicates), free-lancers and publicists. Emphasis is on news, and there are no editorials, though there are opinion columns and cartoons.

The Stars and Stripes carries no advertising and supplements its circulation revenue with the sale of books and magazines, and the operation of a printing plant for U.S. military agencies. The newspaper plant in West Germany is one of the most modern in Europe and includes video display terminals (VDTs) and other electronic equipment. The military address is APO New York 09175 and the international address is Postfach 11 10 34, D-6100 Darmstadt 11.

On May 14, 1945, a week after V-E Day, The Stars and Stripes launched a Hawaiian edition, which became the Pacific edition when it moved on October 3, 1945, from the Honolulu Advertiser to The Asahi Shimbon plant in Tokyo.

Pacific Stars and Stripes now is a 24-page daily tabloid published in a four-floor plant at Hardy Barracks in Tokyo for the 195,000 troops in Japan, Korea, Taiwan, The Philippines, Okinawa, Micronesia and elsewhere in the Pacific. The current paid circulation (at 25 cents a copy) of about 40,000 (as compared to 268,000 during the Viet Nam war) and other revenues produce an annual income of $9,500,000. However, Navy Captain Heber Darton, editor-in-chief, and Army Major Al J. Monica, deputy editor-in-chief, note that P S & S is like other daily newspapers in terms of business problems. The current annual deficit is made up by the U.S. Adjutant General.

In 1946, two Stars and Stripes staff writers, Oram ("Bud") Hutton and Andy Rooney wrote a loving reminiscence, "The Story of The Stars and Stripes" (Farrar & Rinehart). Here's a part of the nostalgic conclusion:

"Here and there across the newspaper empire, which a bunch of soldiers built, a press slows and the last copy of the local S&S has been printed. The staffers go to a new job, or maybe home, and the city room returns to the kind of normalcy it knew before the pinwheels came."

In 1978, Andy Rooney (now a writer and broadcaster at CBS-TV) reminisced about The Stars and Stripes.

"In its prime, it was a very good and professionally written newspaper. We operated under a protective directive from Eisenhower that gave us a lot of freedom under the circumstances. Every colonel with a battalion and every general with a division wanted to tell us what to say about his unit. We resisted, and were protected against recrimination in almost all cases.

"In later years, after the war, The Stars and Stripes became something quite different. The change probably had more to do with the enlisted men than with the officers because, as we used to say, in peacetime the enlisted men are no smarter than the officers."

* * *

The production of this directory is a private enterprise. Research assistance was conscientiously provided by Camille Colon. The book has not been authorized, subsidized or associated in any way with any organization other than the publisher. As with all directories, particularly a first edition, there undoubtedly are errors. The author requests corrections, additions and comments, and hopes that there will be sufficient interest to warrant a second edition.

Guide to Use of the Directory

U.S. listings are by state. Overseas listings are by Air Post Office (APO) and Fleet Post Office (FPO). The FPO's are Navy and Marine publications, with a few exceptions.

Publications which accept advertising are indicated with an asterisk (*).

Most of the official publications do not accept advertising and generally are not interested in publicity from civilian sources. Most of the non-local news comes from military organizations. The editorial staff of an unofficial publication generally consists of military personnel and the editorial office is at the military unit. In these cases, advertising is handled by a civilian enterprise, generally a nearby civilian weekly newspaper. A publication which has a separate advertising office has this information at the end of its listing.

Some of the publications, including many large-circulation newspapers and magazines, are completely private commercial enterprises and are as receptive to publicity material as any civilian publication.

Page size generally is stated in inches, by width and length. For example, 8 x 10 is 8 inches wide and 10 inches deep. Occasionally, particularly where fractions are used, the figures refer to actual printing area. In a few publications, the size is stated in units other than inches, such as the metric system or printing specifications.

One centimeter (cm) is .39 inch and one millimeter is .04 inch.

Column width occasionally is expressed in nonpareils (nonps) or picas. A nonps is 1/12 and pica is 1/6 inch.

Column length occasionally is stated in agate lines. 14 agate lines equal one inch.

A tabloid newspaper, which is smaller than a full-sized newspaper, often is 11 x 14, as compared to the full size of 14 x 23.

Newspaper advertising rates generally are for an agate line, or occasionally, one inch of one column. Magazine rates generally are for a full-page black-

and-white advertisement.

Advertising rates vary according to size, frequency, use of color and other factors. Ad rates in this directory are a general guide. Advertisers should request rate cards from publications for current charges, mechanical requirements, deadlines and other specifications.

Many advertisers and advertising agencies, particularly national brands, utilize Military Media Inc., an extremely efficient "one-step insertion servicice" which represents over 175 military base newspapers in the U.S. and overseas.

Formed in 1974, Military Media Inc. is headed by John and Judith Bradbury, a dynamic couple with considerable military and advertising experience. The company, located at 257 Central Park West, N.Y. 10024 (phone: 212/724-6707), provides unique counseling about military advertising and also handles all details from insertion orders to tear sheets. There is no charge to the advertiser, since Military Media is paid a sales commission by the publishers.

Born in Greenwich, Connecticut, a few months before the start of the Second World War, John Bradbury was graduated from Phillips Exeter Academy and Harvard College. He and his wife live in a pleasant apartment overlooking Central Park. Commuting is rather easy, since their office is in the same building, five floors above their apartment.

As important as the ease in media placement is the marketing expertise that is available to clients of Military Media. Such questions as "Will my product or service be needed or wanted in the military?," "When is the best time to advertise?," "How much vacation time do military personnel receive?" and "Just how much of a youth market is the military?," can be answered by the Bradburys.

Military Abbreviations

ABG	Aberdeen Proving Ground
Abn	Airborne
AFB	Air Force Base
AMC	Army Medical Center
APO	Army Post Office
AVN	Autovon
Bn	Battalion
FPO	Federal Post Office
Inf. Div.	Infantry Division
MCB	Marine Corps Base
MCRD	Marine Corps Recruit Depot
MTMC	Military Traffic Management Command
NAR	Naval Air Reserve
NAS	Naval Air Station
NCOA	National Commissioned Officers Association
NCOA	Non-Commissioned Officer Academy
NSA	Naval Support Activity
OI	Office of Information
PAO	Public Affairs Office
SAC	Strategic Air Command
TAW	Tactical Airlift Wing
TRADOC	Training and Doctrine Command
USAF	U.S. Air Force
USAFA	U.S. Air Force Academy
USCGA	U.S. Coast Guard Academy

Note: The two major Army commands are TRADOC (Training and Doctrine Command) and FORSCOM (Forces Command). Within these commands are units for training (first letter is a T) and for fighting and support.

Major Publications (in order of circulation)

American Legion	New York, N.Y.	Monthly	2,629,000
V.F.W. Magazine	Kansas City, Mo.	Monthly	1,664,000
National News of American Legion Auxiliary	Indianapolis, Ind.	Bimonthly	851,000
Army Reserve Magazine	Washington, D.C.	Bimonthly	585,000
DAV Magazine	Cincinnati, Ohio	Monthly	580,000
V.F.W. Auxiliary	Kansas City, Mo.	10x/year	560,000
Ladycom	Washington, D.C.	8x/year	400,000
Retired Naval Personnel	Washington, D.C.	Quarterly	350,000
Retired Officer	Washington, D.C.	Monthly	260,000
SAM	Arlington, Va.	Monthly	236,000
Family	New York, N.Y.	10x/year	230,000
Soldiers	Alexandria, Va.	Monthly	210,000
Air Reservist	Washington, D.C.	Monthly	200,000
Driver	Norton AFB, Ca.	Monthly	200,000
Army Times	Washington, D.C.	Weekly	142,000
Air Force Magazine	Washington, D.C.	Monthly	138,000
Naval Affairs	Washington, D.C.	Monthly	135,000
Air Force Times	Washington, D.C.	Weekly	134,000
NCOA Journal	San Antonio, Tx.	Monthly	132,000
Airman	Washington, D.C.	Monthly	131,000
Minnesota Legionnaire	St. Paul, Minn.	Biweekly	122,000
Hoosier Legionnaire	Indianapolis, Ind.	9x/year	120,000
California Legionnaire	San Francisco, Ca.	Monthly	119,000
Continental Marine	New Orleans, La.	Monthly	117,000
The Stars and Stripes	A.P.O., N.Y.	Daily	110,000
National Amvet	Washington, D.C.	Bimonthly	110,000
The Officer	Washington, D.C.	Monthly	108,000
Jewish Veteran	Washington, D.C.	Bimonthly	105,000
Navy Times	Washington, D.C.	Weekly	97,000
Iowa Legionnaire	Des Moines, Iowa	Monthly	96,000
Armed Forces Network TV Guide	Frankfurt, Germany	Monthly	92,000
All Hands	Washington, D.C.	Monthly	90,000
Army	Washington, D.C.	Monthly	86,000

Alabama

*McClellan News, Public Affairs Office, Fort McClellan, AL 36205, (205) 238-5574, Weekly, 6,000, 16 pages, 8 x 21, $.25/line, Adv. to Oxford Sun Publishers, Box 3388, Oxford, AL 36203.

Military Police Law Enforcement Journal, U.S. Army MP School/Training Center, Fort McClellan, AL 36205, (205) 238-5121, Quarterly, 10,000, 64 pages, 8½ x 11.

*The Army Flier, Public Affairs Office, Bldg. 121, Fifth Avenue, Fort Rucker, AL 36362, (205) 255-4117, Weekly, 10,000, 16 pages, 8 x 21½, $.22/line, Adv. to Enterprise Ledger, Inc., (205) 347-9533, Box 1140, Enterprise, AL 36330.

U.S. Army Aviation Digest, Drawer P, Ft. Rucker, AL 36362, (205) 255-3619, Monthly, 30,000, 48 pages, 8 x 10½.

Extension Course Institute Catalog and Guide, Extension Course Institute (ECI), Gunter Air Force Station, AL 36118, (205) 279-4258, Monthly.

*Redstone Rocket, Redstone Arsenel, Huntsville, AL 35805, (205) 533-2703, Weekly, 15,000, Tabloid, $.20/line, Adv. to Enquire Printing Co., Box 535T, Huntsville, AL 35805.

*Dispatch, Maxwell AFB, AL 36112, (205) 823-7427, Weekly, 10,800, Tabloid, $.325/line, Adv. to Box 4099, Montgomery, AL 36101.

AFROTC Education Journal, HQ. AFROTC/SBCC, Maxwell AFB, AL 36112, (205) 293-5270, 3 times/year, 2,300, 60 pages, 8 x 10½.

***Publications marked with an asterisk accept advertising. Among the newspapers which accept advertising, most are represented by Military Media, Inc., 257 Central Park West, N.Y. 10024.**

The Air Force Law Review, Air Force Judge Advocate General School, Maxwell AFB, AL 36112, (205) 293-6545, Quarterly, 4,000, 110 pages, 5 x 7½.

Air University Review, Bldg. 1211, Maxwell AFB, AL 36112, (205) 293-2773, Bimonthly, 20,000, 96 pages, 7⅜ x 9¾.

Alaska

Eielson Air Force Base Goldpanner, 5010th Combat Support Group Information Division, Eielson AFB, AK 99702, (907) 377-3107, Weekly, 7,000, 4 pages, 10 x 15½.

Sourdough Sentinel, Elmendorf AFB, AK 99506, (907) 752-2493, Weekly, 6,000, 20 pages, 11 x 17.

Alaskan Pioneer, Public Affairs Office, 172nd Infantry Brigade (AK), Fort Richardson, AK 99505, (907) 862-9298, Weekly, 4,000, 12 pages, 9½ x 15½.

Yukon Sentinel, Public Affairs Office, Fort Wainwright, AK 99703, (907) 353-7117, Weekly, 3,700, 8 pages.

Arizona

Accent, Commander, U.S. Army Communications Command, Fort Huachuca, AZ 85613, (602) 538-2684, Monthly, 12,000, 20 pages, 10¼ x 14.

*Huachuca Scout, Headquarters, Ft. Huachuca, AZ 85613, (602) 538-3418, Weekly, 7,200, 16 pages 12½ x 21, $.215/line, Adv. to Jim Watry, P.A.O. Ft. Huachuca, AZ 85613.

Military Intelligence Magazine, Fort Huachuca, AZ 85613, (602) 538-3033, Quarterly, 6,000, 64 pages, 8¼ x 11.

Civilian Personnel Newsletter and Supervisor's Courier, Luke AFB, 58 Combat Support Group/DPC, Luke AFB, AZ 85309, (602) 935-7411, Monthly, 600, 4 pages, 8 x 10½.

*Luke AFB Tallyho, Box 9, Peoria, AZ, 85345, (602) 979-5894, Weekly, 7,500, 24 pages, 59 pica x 16", $.25/line, Adv. to Bud Bates, Box 9, Peoria, AZ 85345.

Navy Mothers' News, Navy Mothers' Clubs of America, 2126 W. Marshall Avenue, Phoenix AZ 85015, Monthly, 10,000.

*Desert Airman, Davis-Monthan AFB, Tucson, AZ 85707, (602) 748-3204, Weekly, 8,000, 20 pages, $.37/line, Adv. to Territorial Publishers, 1 W. Orange Grove Road, Tucson, AZ 85704.

*Jet Gazette, Williams AFB, AZ 85224, (602) 988-2611, 4,000, Tabloid, $.19/line, Weekly, Adv. to 115 Williams Field Rd., Chandler, AZ 85224.

*Cactus Comet, Public Affairs Office, Marine Corps Air Station, Yuma, AZ 85364, (602) 726-2464, Weekly, 5,000, 8 pages, 8 x 21, $.134/line.

Arkansas

Blythe Spirit, Blytheville AFB, AR 72315, (501) 763-4461, Weekly, 3,800, 12 pages, Tabloid, $.12/line, Adv. to Courier News, Broadway and Moultric, Blytheville, AR 72315.

Arkansas

*Blythe Spirit, Blytheville AFB, AR 72315, (501) 763-4461, Weekly, 3,800, 12 pages, Tabloid, $.12/line, Adv. to Courier News, Broadway and Moultric, Blytheville, AR 72315.

*Air Scoop, 314 Tac Alft Wg, Little Rock AFB, AR 72076, (501) 988-3601, Weekly, 6,500, 12 pages, 15 x 23, $.28 line, Adv. to Canfield Publishing, 116 W. Hickory, Jacksonville, AR 72076.

California

Volunteer, USASFDRC, Bldg. 3, Federal Center, Alameda, CA 94501, (415) 273-7607, Monthly, 300, 8 pages, 8 x 14.

Prospector, Commanding General (B130), Marine Corps Logistics Support Base, Pacific, Barstow, CA 92311, (714) 577-6356, Weekly, 2,500, 8 pages, 11½ x 15.

Tiger Rag, Flying Tigers, c/o Lt. Col. Donald L. Rodewald, 1220 N. Fifth Street, Burbank, CA 91504, (213) 845-3973, Quarterly, 250.

*Desert Wings, Drawer 2425, California City, CA 93515, (714) 373-8686, Weekly, 6,000, Tabloid, $.185/line, Adv. to Enterprise Publishing Co., Drawer 2425, California City, CA 93515.

Benchmark, Employee Relations & Services Branch, Civilian Personnel Office, Building 2265, Marine Corps Base, Camp Pendleton, CA 92055, (714) 725-4434, Biweekly, 2,525, 2 pages, 8 x 14½.

The Pendleton Scout, Joint Public Affairs Office, MCB Camp Pendleton, CA 92055, (714) 725-5202, Weekly, 15,000, 12 pages, Tabloid.

*Valley Bomber, 93D Bombardment Wing, Office of Information, Castle AFB, CA 95340, (209) 726-2878, Weekly, 16 pages, $.176/line, Adv. to Merced-Sun Star, Merced, CA 95340.

The Rocketeer, Naval Weapons Center, China Lake, CA 93555, (714) 939-3354, Weekly, 9,000, 8 pages, 11 x 15.

Distaff Side, Com. Nav. Air Pac., Coronado, CA 92118, (714) 274-4519, Quarterly, 20 pages.

Bulletin, National Order of Trench Rats, 10224 La Reina Avenue, Downey, CA 90241, (213) 923-7888, 5 times/year, 5,000.

The Sandpaper, National Porachito Test Range, El Centro, CA 92243, (714) 339-2556, Weekly, 600, 6 pages, 15 x 11.

*North Islander, Northcoast Publications, Box 878, Encinitas, CA 92024, (714) 753-6543, Weekly, 15,000, Tabloid, $.50/line, Adv. to Northcoast Publications, Box 878, Encinitas, CA 92024.

*Shore Leaf, Northcoast Publications, Box 878, Encinitas, CA 92024, (714)753-6543, Weekly, 12,000, Tabloid, $.50/line, Adv. to Northcoast Publications, Box 878, Encinitas, CA 92024.

*Tailwind, Travis AFB, Box 47, Fairfield, CA 94533, (707) 425-4646, Weekly, 9,000, 28 pages, 6 cols. x 13½, $.28/line.

The Bellringer, Headquarters, U.S. Army Western Regional Recruiting Command, Fort Baker, CA 94129, (415) 561-7330, Monthly, 1,500, 8 pages, 8½ x 10.

Bayonet, 1st Brigade, 7th Inf. Div., Fort Ord, CA 93941, Monthly, 650, 12 pages, 6½ x 8.

14th Engineer Battalion Dimension, 14th Engineer Battalion, Fort Ord, CA 93941, (408) 242-4771, Monthly, 16 pages, 8 x 8.

*Hi-Desert Flyer, Information Office, George AFB, CA 92392, (714) 269-2800, Weekly, 3,500, 12 pages, 14 x 22, $.16/line, Adv. to Daily Press, P.O. Drawer AA, Victorville, CA 92392.

The Communicator, Sierra Army Depot, Herlong, CA 96113, (916) 827-9313, Biweekly, 1,300, 4 pages, 11¼ x 15.

International Defense Review, Interavia Publications, 1633 Vista del Mar, Hollywood, CA 90028, (213) 461-4981, 9 times/year, 26,000, 175 pages, 9¼ x 12¾.

*The Flight Jacket, 23011 Moulton Parkway, Laguna Hills, CA 92653, (714) 581-8100, Weekly, 12,500, 5 cols. x 16, Adv. to Earl F. Shelley, 23011 Moulton Parkway, Laguna Hills, CA 92653.

Family Housing Newsletter, NAS, Lemoore, CA 93245, (209) 998-3492, 1,700, 1 page, 8 x 12.

*The Golden Eagle, Public Affairs Office, NAS, Lemoore, CA 92345, (209) 998-3140, Weekly, 6,000, 12 pages, 14½ x 23, $.18/line, Adv. to Coalinga Record, Box 496, Coalinga, CA 93210.

*Mesa Missilier, 115 No. H St., Box 578, Lompoc, CA 93436, (805) 736-2313, Weekly, 6,263, 16 pages, 8 x 21, $.22/line, Adv. to Robert Bostick, Adv. Mgr., Lompoc Record Publications, Box 578, Lompoc, CA 93436.

The Digest, Long Beach Naval Shipyard, Long Beach, CA 90822, (213) 547-6317, Weekly, 8,000, 4 pages, 11½ x 17½.

Personnel Memo, Box 92960, Attn.: HQ 6592/DPCD, Los Angeles, CA 90009, (213) 643-2237, Monthly, 4 pages, 8 x 10½.

The Navigator Magazine, 323 FTW/DOTN, Mather AFB, CA 95655, (916) 364-2138, Quarterly, 32 pages, 8½ x 11.

*The Classmate, U.S. Naval Postgraduate School, Monterey, CA 93940, 10 issues annually, 1,800, 32 pages, 8 x 10, $7/col. in.

*Off Duty, 4101 Westerly Place, Newport Beach, CA 92660, (714) 833-1633, monthly, 390,000, 38 pages, 8½ x 11, $8855/page, Adv. to The Rios Group, 250 East 63 Street, New York, NY 10021.

Aerospace Safety, Norton AFB, CA 92409, (714) 382-2633, Monthly, 25,000, 32 pages, 8¼ x 11¼.

Driver, AFISC/SEDD, Norton AFB, CA 92409, (714) 382-5675, Monthly, 200,000, 32 pages, 8¼ x 10¾.

Maintenance Magazine, Norton AFB, CA 92409, (714) 382-2113, Quarterly, 25,000, 64 pages, 5 x 7.

The Acorn, Naval Supply Center, Oakland, CA 94625, (415) 466-5202, Semimonthly, 2,400, 8 pages, 8½ x 11.

Navy Family Housing News, Editor, Navy Public Works Center, San Francisco Bay, Box 24003, Oakland, CA 94623, (415) 466-3655, Monthly, 3,900, 8 pages, 8½ x 11.

PWC Baygram, Navy Public Works Center, San Francisco Bay, Box 24003, Oakland, CA 94623, (415) 466-3655, Semimonthly, 1,400, 8 pages, 8½ x 11.

Western Arrow, Public Affairs Office, Oakland Army Base, CA 94626, (415) 466-3021, Bimonthly, 1,500, 4 pages.

The Channel, Women's Army Corps—Veterans Association, c/o Ruth Stubbs, 3839-37 Vista Campana S., Oceanside, CA 92054, (714) 757-8959, Monthly, 2,000.

*Fort Ord Panorama, Box 508, 305 Forest Avenue, Pacific Grove, CA 93950, (408) 649-1311, Weekly, 17,000, 14 pages, Broadsheet, $.30/line, Adv. to C.R. Lucas, Box 508, 305 Forest Ave., Pacific Grove, CA 93950.

The Missile, Public Affairs Office, Pacific Missile Test Center, Pt. Mugu, CA 93042, (805) 982-8094, Weekly, 9,500, 12 pages, 10 x 13.

Naval Air Reserve News, Naval Air Reserve Unit, Naval Air Station, Point Mugu, CA 93042, (805) 982-7296, Monthly, 7,000, 4 pages, 11½ x 14.

The Naval Civil Engineer, Naval School, Civil Engineer Corps Officers, Port Hueneme, CA 93043, (805) 982-3104, Quarterly, 14,500, 48 pages, 8½ x 11.

Seabee Coverall Newspaper, Naval Construction Battalion Center, Port Hueneme, CA 93043, Biweekly, 10,000, 8 pages, 11 x 15.

*Mather AFB Wing Tips, 11379 Pyrites Way, Rancho Cordova, CA 95670, (916) 635-3300, Weekly, 6,600, 20 pages, 6 col. x 13½, $.18/line, Adv. to Wing Tips, Box 38, Fair Oaks, CA 95628.

*McClellan AFB Spacemaker, 11379 Pyrites Way, Rancho Cordova, CA 95670, (916) 635-3300, Weekly, 18,130, 20 pages, Tabloid, $.251/line, Adv. to Spacemaker, Box 38, Fair Oaks, CA 95628.

*Globetrotter, Redlands Daily Facts, Box 191, Redlands, CA 92393, (714) 793-3221, Weekly, 8,500, Standard, $.22/line, Adv. to Redlands Daily Facts, Box 191, Redlands, CA 92373.

*Beacon, Defense Publications, Box 792, Riverside, CA 92501, (714) 686-5585, Weekly, 6,500, Standard $.26/line, Adv. to Defense Publications, Box 792, Riverside, CA 92501.

The Intercom, Sacramento Army Depot, 8350 Fruitridge Road, Sacramento, CA 95813, (415) 421-2324, Monthly, 3,500, 8 pages, 11½ x 15.

Westword, Western Division, Naval Facilities Engineering Command, Box 727, San Bruno, CA 94066, (415) 871-6600, Ext. 2924, Monthly, 1,100, 10 pages, 8 x 10½.

ABA Newsletter, American Battleship Association, Box 11247, San Diego, CA 92111, (714) 271-6106, Semiannual, 2,000.

The Centerline, Public Affairs Office, Navy Public Works Center, San Diego, CA 92136, (714) 235-2155, Monthly, 2,500, 16 pages, 8 x 10.

The Chevron, Public Affairs Office, MCRD San Diego, CA 92140, (714) 2253141, Weekly, 8 pages.

Comnavairpac GSE Newsletter, Comnavairpac Code 75M, Naval Air Station, North Island, San Diego, CA 92135, (714) 437-6101, Bimonthly, 1,000, 16 pages, 8 x 10½.

DATC FMAG "DIALOG," Box 116, NAVSTA, San Diego, CA 92135, (714) 235-1771, Monthly, 1,400, 8 pages, 8½ x 11.

*The Dry Dock, Naval Regional Medical Center, Bldg. 153-1, San Diego, CA 92134, (714) 233-2662, Weekly, 10,000, 12 pages, 10¼ x 16, $.50/line, Adv. to Peggy True, North Coast Printers, 687 First Avenue, Encinitas, CA 92024.

The Hoist, Naval Training Center, San Diego, CA 92133, (714) 225-5311, Weekly, 7,000, 8 pages, 11 x 16.

*The Jet Journal, Naval Air Station Miramar, San Diego, CA 92145, (714) 271-3513, Weekly, 7,000, 8 pages, 10 x 16, $.30/line.

*Long Beach Dispatch, Western State Weeklies, Inc., 6312 Riverdale St., San Diego, CA 92120, (714) 280-2985, Weekly, 10,000, Tabloid, $.35/line.

Narfacts, Naval Air Rework Facility, Naval Air Station, Code 120, Bldg. 94, San Diego, CA 92135, (714) 437-5791, Bimonthly, 6,500, 4 pages, 8 x 10½.

*Navy News, Box 4528, 2228 El Cajon Blvd., San Diego, CA 92104, (714) 2912131, Monthly, 15,000, 16 pages, 6 x 16, $4.50 to $7/col. in.

Outlook, Naval Ocean Systems Center, Code 032, 271 Catalina Blvd., San Diego, CA 92152, (714) 225-6238, Bimonthly, 500, 8 pages, 17½ x 11½.

The Periscope, Commander Submarine Force, U.S. Pacific Fleet Representative, FSPO, San Diego, CA 92132, (714) 225-7350, Biweekly, 3,000, 4 pages.

*San Diego Dispatch, Western State Weeklies, Inc., 6312 Riverdale Street, San Diego, CA 92120, (714) 280-2985, Weekly, 26,000, Tabloid, $.40/line.

The Sou'wester, 960 N. Harbor Drive, San Diego, CA 92132, (714) 235-3431, Bimonthly, 1,000, 4 pages, 17½ x 11½.

The Triad, Naval Air Reserve Unit, Internal Relations Office, NAS, North Island, San Diego, CA 92135, (714) 437-5006, Monthly, 2,500, 6 pages, 11 x 16.

Fog Horn, Letterman Army Medical Center, Presidio of San Francisco, CA 94129, (415) 561-3160, Semimonthly, 5,000, 20 pages, 6 x 9.

Masthead, Bldg. 1, Naval Support Activity, Treasure Island, San Francisco, CA 94130, (415) 765-5505, Biweekly, 3,000, 8 pages, 11½ x 16½.

Star Presidian, Headquarters Presidio, San Francisco, CA 94132, (415) 561-2005, Weekly, 15,000, 4 pages, Tabloid.

*The Moffett News, Box 60783, Sunnyvale, CA 94088, (415) 966-5976, Weekly, 8,000, 12 pages, 11 x 17.

*Sharpshooter, Box 419, Tracy, CA 95376, (209) 948-1131, Monthly, 7,000, Tabloid, $.19/line, Adv. to Tracy Press, Box 419, Tracy, CA 95376.

CIVINFO, Civilian Personnel Office, Marine Corps Base, Bldg. 1042, Twentynine Palms, CA 92278, (714) 367-9111, Weekly, 900, 4 pages, 8½ x 11.

*Desert Trail, Box 159, 29 Palms, CA 92277, (714) 367-3577, Weekly, 10,000, Standard, $.21/line, Adv. to Box 159, 29 Palms, CA 92277.

The Observation Post, Marine Corps Base, Twentynine Palms, CA 92278, (714) 367-9111, Weekly, 5,000, 12 pages, 11½ x 15.

Mare's Tale, Naval Support Activity, Mare Island, Vallejo, CA 94590, (707) 646-4465, Semimonthly, 4,200, 8 pages, 11 x 17.

*The Space Sentinel, 430 Center Street, Yuba City, CA 95991, (916) 673-9414, Weekly, 6,500, 16 pages, 8 x 21, $.18/line, Adv. to Louise M. Tweith, Box 574, Yuba City, CA 95991.

Colorado

*Lowry Airman, Sentinel Newspapers, 11059 East Bethany Drive, Aurora, CO 80014, (303) 750-7555, Weekly, 14,500, $.50/line, Adv. to Sentinel Newspapers, 11059 East Bethany Dr., Aurora, CO 80014.

ADCOM Chaplain Newsletter, HQ ADCOM/HCX, Peterson AFB, CO 80916, (303) 635-8911, Quarterly, 400, 12 pages, 8 x 10½.

Crossfeed, 7300 East First Avenue, Denver, CO 80280, (303) 394-4901, Quarterly, 1,700, 7 pages, 8½ x 11.

*Mountaineer, Public Affairs Office, Fort Carson, CO 80913, (303) 597-4143, Weekly, 22,000, 20 pages, Tabloid, $.25/line, Adv. to Dorothy Walnutt, Gowdy-Printcraft Press, Inc., 22 N. Sierra Madre, Colorado Springs, CO 80901.

*Observer, Peterson Field, AFB, Norad, CO (303) 634-1593, Weekly, 5,500, 32 pages, 6 col. x 16, $.20/line, Adv. to Dorothy Walnutt, Gowdy-Printcraft Press, Inc., Box 789, Colorado Springs, CO 80901.

Communications, Electronics & Computer Resources Digest, Aerospace Defense Command (USAF), Peterson AFB, CO 80914, (303) 591-7227, Quarterly, 1,700, 24 pages, 8 x 10½.

Interceptor Magazine, Peterson AFB, CO 80914, (303) 635-8911, Monthly, 4,450, 32 pages, 8 x 10½.

*Falconews, USAF Academy, CO 80840 (303) 472-2044, Weekly, 8,000, 12 pages, 12 x 16, $.20/line, Adv. to Dorothy Walnutt, Gowdy-Printcraft, 22 N. Sierra Madre, Colorado Springs, CO 80901.

U.S. Air Force Academy Catalog, Registrar's Office, USAF Academy, CO 80840, (303) 472-2640, 130,000, 160 pages, 8 x 10.

Talon, Box 6066, U.S. Air Force Academy, CO 80840, (303) 472-4644, Monthly, 5,600, 32 pages, 8½ x 11, Adv. to Box 6066, USAFA, CO 80840.

Connecticut

*The Dolphin, Box 44, Naval Submarine Base, New London, Groton, CT 06340, (203) 449-3514, Weekly, 50,000, 16 pages, Tabloid, $.38/line, Adv. to Norwich Bulletin Co., 66 Franklin Street, Norwich, CT 06360.

*The Howling Gale, Box A-37, Chase Hall, U.S.C.G.A., New London, CT 06320, (203) 443-8463, Monthly, 2,000, 24 pages, 8½ x 11, $.20/line.

The Torch, Italian American War Veterans of the United States, 369 S. Leonard, Waterbury, CT 06708, (205) 574-3872, Quarterly, 8,500.

Delaware

*Airlifter, Office of Information, Dover AFB, DE 19901, (302) 678-6881, Weekly, 9,000, 20 pages, 5 col. x 14, $.16/line, Adv. to Dover Post, Inc., Treadway Towers, Dover, DE 19901.

District of Columbia

Air Force Comptroller, HQ USAF, Washington, D.C. 20330, Quarterly, 5,000, 48 pages, 8½ x 11.

*Air Force Magazine, Air Force Association, 1750 Pennsylvania Ave., N.W., Washington, D.C. 20006, (202) 637-3330, Monthly, 141,656, 7 x 10, $2,870/page.

*Air Force Times, 475 School Street, W., Washington, D.C. 20024, (202) 554-7145, Weekly, 134,443, $2,745/page., Adv. to Army Times Publishing Co., 475 School Street W., Washington, D.C. 20024.

Air Reservist, Box 423, Bolling AFB, Washington, D.C. 20332, Monthly, 200,000.

Airman, P.O. Box 423, Bolling AFB, Washington, D.C. 20332, Monthly, 130,750. **(Moved to Kelly AFB, TX 78241.)**

All Hands Magazine, Print Media Division, Room 1044, Crystal Plaza No. 6, Washington, D.C., 20360, (202) OX 2-2564, Monthly, 90,000, 48 pages, 8½ x 10⅞.

American War Mother, American War Mothers, 2615 Woodley Place, N.W., Washington, D.C. 20008, (202) 462-2791, Bimonthly, 11,000.

*Armed Forces Journal International, 1414 22nd Street, N.W., Washington, D.C. 20037, (202) 296-0450, Monthly, 18,500, 36 pages, 8¼ x 11, $1,450/page.

*Army, 1529 18th Street, N.W., Washington, D.C. 20036, (202) 483-1800, Monthly, 86,217, $1,800/page, Adv. to Assn. of U.S. Army, Publisher, 1529 18th Street, N.W., Washington, D.C. 20036.

Army Host, HQDA (DAAG-CM), Washington, D.C. 20314, (202) 693-1059, 2,000, 16 pages, 8 x 10½.

Army Museum Newsletter, U.S. Army Center of Military History, Forrestal Bldg., Washington, D.C. 20314, (202) 693-5041, Quarterly, 1,200, 48 pages, 8½ x 11.

Army Personnel Letter, DAPE-ZX, Washington, D.C. 20310, (202) 697-6314, Monthly, 125,000, 6 pages, 8 x 10.

Army Reserve Magazine, The Pentagon, Room 1E456, Washington, D.C., 20310, (202) 697-2470, Bimonthly, 585,000, 36 pages, 8 x 10.

*Army Times, 475 School Street, W., Washington, D.C. 20024, (202) 554-7145, Weekly, 142,331, $2,745/page.

*Bolling Beam, 1100th ABW/OI, Bolling AFB, D.C., 20332, (202) 767-4781, Weekly, 10,000, 16 pages, Tabloid, $.30/line, Adv. to Sentinel Newspapers, Box 1272, Rockville, MD 20850.

Bulletin, American Veterans Committee, 1333 Connecticut Avenue, N.W., Washington, D.C. 20036, (202) 293-4890, Quarterly, 10,000.

Bulletin, Disabled Officers Association, 1612 K Street, N.W., Washington, D.C. 20006 (202) 331-1034, 9 times/year, 6,000.

Bulletin, Jewish War Veterans of the U.S.A.—National Ladies Auxiliary, 1712 New Hampshire Avenue, N.W., Washington, D.C. 20009, (202) 667-9061, Quarterly.

Careergram, Department of the Navy, Bureau of Naval Personnel, Washington, D.C. 20370, (202) 692-8230, Bimonthly, 7,500, 4 pages, 8½ x 10.

Carry On, Women's Overseas Service League, Box 39033, Washington, D.C. 20016, Quarterly, 1,700.

*Catholic War Veterans of the U.S.A., Two Massachusetts Avenue, N.W., Washington, D.C. 20001, (202) 737-9600, Bimonthly, 50,000.

Chaplain Corps Items of Interest, Chief of Chaplains, Bureau of Naval Personnel, Department of the Navy, Washington, D.C. 20370, (202) 694-2903, Monthly, 2,500, 12-24 pages, 8 x 10.

*Club Executive, 1028 Connecticut Avenue, N.W., Washington, D.C. 20036, (202) 296-4514, Monthly, 6,304, $580/page, Adv. to R & W Management Co. Inc., 1028 Connecticut Avenue, N.W., Washington, D.C. 20036.

Deckplate—Steam Propulsion Newsletter, Naval Sea Systems Command, Washington, D.C. 20362, (202) 692-0751, Quarterly, 7,000, 12 pages, 8 x 10½.

*Defense Transportation Journal, 1612 K Street, N.W., Washington, D.C. 20006, (202) 331-1414, Bimonthly, 11,500, $615/page.

Direction Magazine, Crystal Plaza 6, Room 1044, Washington, D.C. 20360, (202) 692-2564, Bimonthly, 8,500, 36 pages, 7¾ x 10.

*EEO Newsletter, Headquarters Naval Material Command, Washington, D.C. 20360, (202) 692-2893, Bimonthly, 8 pages, 8 x 10½.

Faceplate, Supervisor of Diving, Naval Sea Systems Command Washington, D.C. 20362, (202) 697-7606, Quarterly, 5,500, 36 pages, 8 x 10.

Fortitudine, History and Museums Division, Headquarters, U.S. Marine Corps, Washington, D.C. 20380, (202) 433-3841, Quarterly, 9,000, 24 pages, 8 x 10½.

*Fridaygram, Department of the Army, Office, Chief of Public Affairs, Attn.: SAPA-ZXP, Washington, D.C. 20310, (202) 695-3405, Bimonthly, 12 pages, 8 x 10½.

Gold Star Mothers, American Gold Star Mothers, 2128 Leroy Place, N.W., Washington, D.C. 20008, (202) 265-0991, Bimonthly, 15,000.

Hotline HQMC, Division of Information, Headquarters Marine Corps, Washington, D.C. 20380, (202) 694-1492, Monthly, 50,000, 12 pages, 8 x 10.

The Inspector General Brief, HQ USAF/IGEP, Washington, D.C. 20330, (202) OX 5-3045, Semimonthly, 43,000, 24 pages, 7⅞ x 10¼.

JAG Journal, Office of the Judge Advocate General, Department of the Navy, Washington, D.C. 20370, (202) 694-5656, Semiannual, 9,370, 100 pages, 6 x 9.

The JAG Reporter, 1900 Hair Street, S.W., Washington, D.C. 20324, (202) OX 3-5830, Bimonthly, 3,250, 45 pages, 8½ x 11.

Jewish Veteran, Jewish War Veterans of the U.S.A., 1712 New Hampshire Avenue, N.W., Washington, D.C. 20009, (202) 265-6280, Bimonthly, 105,000.

*Ladycom, 1800 M Street, N.W., Suite 650 South, Washington, D.C. 20036, (202) 872-0960, 8 times/year, 400,000, 56 pages, $5850/page, Adv. to R. Modny, 520 N. Michigan Avenue, Chicago, IL 60611.

Marine Corps Club System Newsletter, Headquarters, U.S. Marine Corps, Code MSMS-2, Washington, D.C. 20380, (202) 694-2728, Biannual, 500, 40 pages, 8½ x 11.

*Military Chaplain, 7758 Wisconsin Avenue, Washington, D.C. 20014, (301) 657-4077, Bimonthly, 2,700, Adv. to Military Chaplains Assn. of the United States, 7758 Wisconsin Avenue, Suite 401, Washington, D.C. 20014.

*The Military Engineer, Suite 905, 740 15th Street, N.W., Washington, D.C. 20005, (202) 638-4010, Bimonthly, 22,000, 80 pages 8½ x 11, $650/page.

*Military Market Magazine, 475 School St., S.W., Washington, D.C. 20024, (202) 554-7180, Monthly, 9,261, $665/page, Adv. to Army Times Publishing Co., 475 School Street, S.W., Washington, D.C. 20024.

NARTOPIX, Public Affairs Office, Naval Air Reserve Unit, Naval Air Facility, Washington, D.C. 20390, (202) 981-6723, Monthly, 2,300, 8 pages.

The National AMVET, AMVETS (American Veterans of World War II, Korea and Vietnam), 1710 Rhode Island Avenue, N.W., Washington, D.C. 20036, (202) 223-9550, Bimonthly, 150,000.

*National Defense Magazine, 740 15th Street, N.W., Washington, D.C. 20005, (202) 437-7250, Bimonthly, 33,000, 80 pages, 8½ x 11, $295/quarter page, Adv. to Mrs. Joyce Selby, 12005 Chesterton Drive, Upper Marlboro, MD 20870.

*Naval Affairs, 1303 New Hampshire Avenue, N.W., Washington, D.C. 20036 (202) 785-2768, Monthly, 135,000, 24 pages, 8½ x 11, $46/col. in.

Naval Affairs Magazine, Fleet Reserve Association, 1303 New Hampshire Avenue, N.W., Washington, D.C. 20036, (202) 785-2768, Monthly, 134,000.

Naval Aviation News, Potomac Annex, Bldg. 6, 23rd and E Streets, N.W., Washington, D.C. 20372, (202) 254-4696, Monthly, 33,000, 44 pages, 7⅞ x 10¼.

*Naval Engineers Journal, The American Society of Naval Engineers, Inc., 1012 14th Street, N.W., Washington, D.C. 20005, (202) 737-0757, Bimonthly, 5,700, 8¼ x 11¼, $20/col. in.

Navy Foodservice, Navy Food Service Systems Office, Washington, D.C. 20374, (202) 433-3093, Quarterly, 5,000, 16 pages, 8 x 10.

Navy Policy Briefs, Navy Internal Relations Activity, Navy Department, Washington, D.C. 20350, (202) 695-5471, Monthly, 10,000, 6 pages, 8 x 11½.

Navy Supply Corps Newsletter, Naval Supply Systems Command, Washington, D.C. 20376, (202) 697-3767, Monthly, 13,000, 60 pages, 8 x 10½.

*Navy Times, 475 School Street, S.W., Washington, D.C. 20024, (202) 554-7150, Weekly, 97,208, 64 pages, Tabloid, $1,755/page, Adv. to Army Times Publishing Co., 475 School Street, S.W., Washington, D.C. 20024.

News-letter, National Society, United States Daughters of 1812, 1461 Rhode Island Avenue, N.W., Washington, D.C. 20005, (202) 332-3181, 3 times/year, 4,500.

Newsletter, AMVETS (American Veterans of World War II, Korea, and Vietnam), 1710 Rhode Island Avenue, N.W., Washington, D.C. 20036, (202) 223-9550, Monthly 150,000.

Newsletter for Retired Marines, Commandant of the Marine Corps (Code MSPA-3), Headquarters, U.S. Marine Corps., Washington, D.C. 20380, (202) 694-1901, Quarterly, 83,000, 8 pages, 8 x 10½.

Newsletter, National Association of Concerned Veterans, 1900 L Street, N.W., Suite 314, Washington, D.C. 20036, (202) 785-2155, Monthly, 25,118.

Newsletters to Chapters, Widows of World War I, 1234 Massachusetts Avenue, N.W., Suite 206, Washington, D.C. 20005, (202) 638-6311, Monthly, 5,522.

*The Officer, One Constitution Avenue, N.E., Washington, D.C. 20002, (202) 547-4300, Monthly, 108,000, 32 pages, 8¼ x 11, $45/col. in.

Officer Personnel Newsletter, Bureau of Naval Personnel, Navy Department, Washington, D.C. 20370, (202) 694-1402, Quarterly, 14,000, 30 pages, 7¾ x 10¼.

Paraplegia News, Paralyzed Veterans of America, 7315 Wisconsin Avenue, Suite 300 W, Washington, D.C. 20014, (301) 652-2135, Monthly, 11,000.

Procurement Newsletter, Commandant of the Marine Corps, Headquarters, Marine Corps, Washington, D.C. 20380, (202) 694-2315, Quarterly, 400 28 pages, 8 x 10½.

Recreational Services Today (Management Bulletin), Bureau of Naval Personnel, Washington, D.C. 20370, (202) 694-3931, Quarterly, 1,500, 12 pages, 8 x 10½.

The Reliance, HQ AFOSI/OI, Forrestal Bldg., Washington, D.C. 20134, (202) 693-0089, Monthly, 1,000, 20 pages.

Report of NRL Progress, Naval Research Laboratory, 4555 Overlook Avenue, S.W., Washington, D.C. 20375, (202) 767-3374, Monthly, 1,800, 50 pages, 8 x 10.

Retired Naval Personnel Newsletter, Retired Personnel Support Section, Bureau of Naval Personnel, Washington, D.C. 20370, (202) 694-1297, Quarterly, 350,000, 16 pages, 7⅞ x 10¼.

*The Retired Officer Magazine, 1625 Eye Street, N.W., Washington, D.C. 20006, (202) 331-1111, Monthly, 260,000, 56 pages, 8½ x 11, $75/col. in.

*The Review, 1000 Connecticut Avenue, Washington, D.C. 20036, Bimonthly, 7,800, $430/page, Adv. to American Logistics Assn., 1000 Connecticut Avenue, N.W., Washington, D.C. 20036.

*Sea Power, 818 18th Street, N.W., Washington, D.C. 20006, (202) 331-7887, Monthly, 53,000, $1,200/page, Adv. to Navy League of the U.S. 818 18th Street, N.W., Washington, D.C. 20006.

Sealift, 4228 Wisconsin Ave., N.W., Washington, D.C. 20016, (202) 282-2808, Monthly, 7,000, 24 pages, 8½ x 11.

Shock and Vibration Digest, Shock and Vibration Information Center, Naval Research Laboratory, Washington, D.C. 20375, (202) 767-2220, Monthly, 1,500, 100 pages, 8 x 10½.

*Stripe, Public Affairs Office, Walter Reed Army Medical Center, Washington, D.C. 20012, (202) 576-2420, Weekly, 7,000, 16 pages, 10 x 14, $.30/line, Adv. to Stripe, c/o Sentinel Newspapers, Comprint Lane, Gaithersburg, MD 20760.

The Surveyor, Chesapeake Division, Naval Facilities Engineering Cmd., Bldg. 108-1, Washington Navy Yard, Washington, D.C. 20374, (202) 433-3122, Monthly, 600, 8 pages, 8½ x 16.

Translog Magazine, Military Traffic Management Command, Washington, D.C. 20315, (202) 756-1116, Monthly, 13,000, 20 pages, 8⅛ x 10⅞.

Triennial Yearbook, General Society of the War of 1812, 1307 New Hampshire Avenue, N.W., Washington, D.C. 20036, (202) 785-2068, 1,427.

U.S. Air Force Policy Letter for Commanders, HQ USAF (SAF/OII), Washington, D.C. 20330, (202) 695-9776, Semimonthly, 82,000, 4 pages, 8 x 10½.

Supplement to the AF Policy Letter for Commanders, HQ USAF (SAF/OII), Washington, D.C. 20330, (202) 695-9776, Monthly, 31,000, 36 pages, 5⅛ x 7¾.

U.S. Army Chaplains Bulletin, Chaplains Division, Bureau of Naval Personnel, Department of the Navy, Washington, D.C. 20370, (202) 694-2903, 3 times/year, 2,300, 28 pages, 8 x 10.

U.S. Navy Medicine, Department of the Navy, Bureau of Medicine and Surgery, Washington, D.C. 20372, (202) 254-4214, Monthly, 17,000, 28 pages, 7⅞ x 10¼.

U.S. Navy Security Assistance Newsletter, Chief of Naval Operations, Navy Department, Washington, D.C. 20350, (202) 695-4449, Semiannual, 400, 30 pages, 8 x 10½.

United States Air Force Medical Service Digest, HG USAF, Washington, D.C. 20314, (202) 693-5747, Bimonthly, 12,000, 40 pages, 8 x 10½.

What's Sup, Naval Supply Systems Command, Washington, D.C. 20376, (202) 697-3795, Weekly, 1,800, 8 pages, 8½ x 10.

War Cry of 1812, General Society of the War of 1812, 1307 New Hampshire Avenue, N.W., Washington, D.C. 20036, (202) 785-2068, 1,427.

Water Spectrum, Office of the Chief of Engineers, Department of the Army, Washington, D.C. 20314, (202) 693-6777, Quarterly, 14,000, 48 pages, 8½ x 11.

Florida

*Fighter Forum, Box 339, Homestead, FL 33030, (305) 247-2321, Weekly, 7,500, Standard, $.283/line, Adv. to South Dade News Leader, Box 339, Homestead, FL 33030.

*Eglin Eagle, Eglin Air Force Base, Fort Walton Beach, FL 32548, (904) 881-6668, Weekly, 8,000, 10 x 16, $.31/line, Adv. to Southern Publishing Co., Box 1232, Pensacola, FL 32596.

*SOF Hi-Lites, Hurlburt Field, FL 32569, (904) 884-7464, Weekly, 4,800, 16 pages, 10 x 16, $.26/line, Adv. to Pensacola Engraving Company, Pensacola, FL 32596.

*Airwinger, Public Affairs Office, NAS, Cecil Field, Jacksonville, FL 32215, (905) 778-6052, Weekly, 16,000, 32 pages, 6 cols. x 15, Adv. to Holt Publishing Co., 4565 Shirley Avenue, Jacksonville, FL 32210.

*Jax Air News, Box 2, NAS, Jacksonville, FL 32212, (905) 772-2415, Weekly, 16,000, 24 pages, $.35/line, Adv. to Holt Publishing, 4565 Shirley Avenue, Jacksonville, FL 32210.

*Sky Anchor, Green Street Graphics, 1028 Truman Avenue, Key West, FL 33040, (305) 296-8532, Biweekly, 5,000, Tabloid, $.235/line, Adv. to Green Street Graphics, 1028 Truman Avenue, Key West, FL 33040.

Giron, Bay of Pigs Veterans Association, 900 S.W. First Street, Room 23, Miami, FL 33130, (305) 324-6353, Monthly, 2,136.

*The Mirror, U.S. Naval Station, Mayport, FL 32228, (904) 246-5226, Weekly, 6,000, 20 pages, 10¼ x 14, $.26/line, Adv. to Beaches Leader, 609 N. 3rd Street, Jacksonville Beaches, FL 32250.

*Whiting Tower, NAS, Whiting Field, Milton, FL 32570, (904) 623-7651, Weekly, 4,500, 12 pages, 10 x 16, $.26/line, Adv. to Pensacola Engraving, Pace and Government, Pensacola, FL 32596.

*Navigator, Box 3428, Orlando, FL 32802, (305) 425-5672, Weekly, 20,000, Tabloid, $.30/line, Adv. to Military Publications, Box 3428, Orlando, FL 32802.

*Photoscope, Box 12338, Panama City, FL 32401, (904) 763-3975, Monthly, 5,000, 16 pa2 pages, 10 x 16, $.26/line, Adv. to Pensacola Engraving, Pace and Government, Pensacola, FL 32596.

*Navigator, Bx 3428, Orlando, FL 32802, (305) 425-5672, Weekly, 20,000, Tabloid, $.30/line, Adv. to Military Publications, Box 3428, Orlando, FL 32802.

*Photoscope, Box 12338, Panama City, FL 32401, (904) 763-3975, Monthly, 5,000, 16 pages, 8½ x 11, $5/col. in., Adv. to Keevco, Box 2423, Panama City, FL 32401.

*Tyndall Guide and Directory, Box 12026, Panama City, FL 32401, (904) 7633975, Yearly, 15,000, 100 pages, 8½ x 11, Adv. to Box 12026, Panama City, FL 32401.

The Underseer, Naval Coastal Systems Laboratory, Panama City, FL 32407, (904) 234-4420, Monthly, 1,500, 6 pages, 11½ x 15.

The Missileer, 6550 ABW/OI, Patrick AFB, FL 32925, (305) 494-5922, Weekly, 8 pages, 11 x 17.

*Gosport (Naval Air Basic Training bases), Box 1232, Pensacola, FL 32596, (904) 438-5421, Weekly, 14,500, 10 x 16, $.38/line.

*Lemot (Air Force Officers' Wives at Eglin Air Force Base and Hurlburt Field), Box 1232, Pensacola, FL 32596, (904) 438-5421, Weekly, 4,500, 10 x 16, $100/page.

Mustin Missile (Mustin Beach Officers' Club), Box 1232, Pensacola, FL 32596, (904) 438-5421, Monthly, 5,000, 10 x 16, $155/page.

Navy Campus Newsletter, Chief of Naval Education and Training Support, Pensacola, FL 32509, (904) 452-1651, Monthly, 4,000, 4 pages, 8 x 10½.

Naval General Library Services Newsletter, CNET Support (N32), Pensacola, FL 32509, (904) 452-1372, Monthly, 800, 4 pages, 8½ x 14.

*Officers' Wives League Bulletin (Naval, Marine and Coast Guard officers' wives in Pensacola area), Box 1232, Pensacola, FL 32596, (904) 438-5421, Monthly, $6,000, 10 x 16, $155/page.

The Airlifter, The Airlift Association, Box 788, Sarasota, FL 33578, (813) 366-6091, Quarterly, 2,000.

*Thunderbolt of MacDill AFB, 815 W. Platt Street, Tampa, FL 33606, (813) 251-1673, Weekly, 6,500, 16 pages, 6 col. x 16, $.34/line, Adv. to J.W. Martin, 815 W. Platt Street, Tampa, FL 33606.

Civilian Courier, 4756 Air Base Group/DPC, Tyndall Air Force Base, FL 32403, (904) 283-6206, Monthly, 350, 8 pages, 8 x 10½.

*Jet Scope, Tyndall AFB, FL 32403, (904) 283-2965, Weekly, 5,000, 12 pages, 11½ x 15, $.195/line, Adv. to 123 W. 5th St., Panama City, FL 32401.

NAFI Notes, 4756 Air Base Group/DPC, Tyndall AFB, FL 32403, (904) 283-6206, Quarterly, 300, 4 pages, 8 x 10½.

USAF Interceptor Weapons Newsletter, USAF Interceptor Weapons School, Tyndall AFB, FL 32403, (904) 283-2824, Quarterly, 350, 40 pages, 8 x 10½.

Georgia

*Knight Express, Moody AFB, GA 31601, (912) 333-4211, Weekly, 3,000, 20 pages, 5 x 14, $.115/line, Adv. to Patten Publications, Adel, GA 31620.

Commissary Newsletter, Marine Corps Logistics Support Base Atlantic, Albany, GA 31704, (912) 435-1721, Monthly, 250, 1 page, 8 x 10½.

*The Emblem, Public Affairs Office, Marine Corps Supply Center, Albany, GA 31704, (912) 439-5216, Weekly, 4,200, 20 pages, 10 x 14, $.286/line.

MCLSBLANT Newsletter, Marine Corps Logistics Support Base, Atlantic, Albany, GA 31704, (912) 439-6571, Quarterly, 2,500, 8 pages, 8 x 10½.

Civilian News Briefs, Navy Supply Corps School, Athens, GA 30606, (404) 549-6620, Monthly, 4 pages, 8 x 10½.

Dixie Digest, Public Affairs Branch, 75 Piedmont Avenue, N.E., Atlanta, GA 30303, Monthly, 2,000, 8 pages.

*The Fort Gordon Rambler, Box 632, Augusta, GA 30903, (404) 724-3627, Weekly, 6,000, 10 pages, 15 x 21, $.40/line.

*The Bayonet, Box 711, Fort Benning, Columbus, GA 31902, (404) 322-8831, Weekly, 16,500, 28 pages, $.37/line, Adv. to Bob Turner, Box 711, Columbus, GA 31902.

Infantry Magazine, Box 2005, Fort Benning, GA 31905 (404) 545-2350, Bimonthly, 25,000, 64 pages, 8¼ x 11.

The Examiner, Eisenhower Medical Center, Fort Gordon, GA 30905, (404) 791-2555, Biweekly, 14,500, 8 pages, Tabloid.

Patriot, HQ, 24th Inf. Div. & Fort Stewart, Fort Stewart, GA 31313, (912) 767-3527, Weekly, 10,000, 16 pages, 11½ x 15.

*Sentinel, Box 368, Fort McPherson, Jonesboro, GA 30237, (404) 471-4742, Weekly, 5,000, 20 pages, 5 cols. x 14, $.35/line.

Personnel Newsletter, HQ AFRES/DP, Robins AFB, GA 31098, (404) 926-6365, Monthly, 375, 15 pages, 8 x 10½.

*Robins Rev-Up, Warner Robins ALC-OI, Robins AFB, GA 31098, (912) 926-2137, Weekly, 12,500, 20 pages, 11½ x 14, $.21/line, Adv. to The Rev-Up, Drawer A., Warner Robins, GA 31093.

Hawaii

The Pointer, NAS Barbers Point, HI 96862, (808) 684-7101, Weekly, 4,468, 8 pages, 16 x 11⅜.

Chief of Supply Newsletter, 15 Supply Squadron/LGSPS, Hickam AFB, HI 96853, (808) 449-6822, Quarterly, 8 pages, 8 x 10½.

15th Air Base Wing Consolidated Base Personnel Newsletter, Hickam AFB, HI 96853, (808) 449-9941, Monthly, 465, 16 pages, 8 x 10½.

Hawaiian Falcon, 15 ABW/OIF, Hickam AFB, HI 96853, (808) 449-6662, Weekly, 5,000, 8 pages, Tabloid.

*Military Sun Press/Navy News, 46-016 Alaloa St., Honolulu, HI 96744, (808) 235-5881, Weekly, 26,800, $.34/line.

Pupukahi, State of Hawaii, Department of Defense, Fort Ruger, Honolulu, HI 96816, (808) 732-1711, Quarterly, 6,500, 8 pages.

Trade Winds, Naval Communications Area Master Station, EASTPAC, Honolulu, Wahiawa, HI 96786, (808) 653-5317, Monthly, 1,000, 8 pages, 14 x 11½.

Aloha Hoalauna, Navy Public Works Center, Pearl Harbor, HI 96860, (808) 474-3149, Monthly, 7,000, 4 pages.

Centergram, Navy Public Works Center, Pearl Harbor, HI 96860, (808) 474-3149, Monthly, 1,500, 4 pages, 11½ x 16.

COMSUBPAC 3-M Newsletter, Commander Submarine Force, U.S. Pacific Fleet, Pearl Harbor, HI 96860, (808) 471-9335, Quarterly, 1,500, 3 pages, 8 x 10½.

*Hawaii Navy News, Naval Station Pearl Harbor, HI 96860, (808) 471-0239, Weekly, 15,000, 16 pages, $.345/line, Adv. to Community Publications, Inc., 46-016 Alaloa Street, Kaneohe, HI 96744.

Pacific Health Bulletin, Navy Environmental and Preventive Medicine Unit No. Six, Box 112, Pearl Harbor, HI 96860, (808) 471-9505, Monthly, 1,100, 4 pages, 8 x 10½.

Tropic Lightning News, Public Affairs Office, 25th Inf. Div. Schofield Barracks, HI 96857, (808) 655-4815, Biweekly, 15,000, 16 pages, 12 x 16.

Caducean, Public Affairs Office, Tripler Army Medical Center, HI 96859, (808) 433-5785, Biweekly, 4,000, 8 pages, 11½ x 16.

Idaho

*Wing Spread, Mtn. Home AFB, ID 83647, (208) 828-6123, Weekly, 4,800, 28 pages, 11½ x 16½, Adv. to Mtn. Home News, Mtn. Home, ID 83647.

Illinois

Can Do, Seabee Veterans of America, 218 Granada Road, Carpentersville, IL 60110, (312) 428-2732, Quarterly, 2,000.

DCASR News, Box 66475, O'Hare International Airport, Chicago, IL 60666, (312) 694-3031, Monthly, 4 pages, 11 x 16.

National PLAV News, Polish Legion of American Veterans, U.S.A., 3024 N. Laramie Avenue, Chicago, IL 60641, (312) 283-9161, Quarterly, 15,000.

Polaris, United States Submarine Veterans of World War II, 862 Chatham Avenue, Elmhurst, IL 60126, (312) 834-2718, Bimonthly, 12,000.

U.S. Army Recruiting and Career Counseling Journal, HQ., Fort Sheridan, IL 60037, (312) 926-3918, Monthly, 20,000, 36 pages, 8½ x 11.

*Fort Sheridan Tower, 30 South Whitney Street, Box 268, Grayslake, IL 60030, (312) 223-8161, Weekly, 7,000, 12 pages, 10 x 15, $.77/line.

*Great Lakes Bulletin, 30 South Whitney Street, Box 268, Grayslake, IL 60030, (312) 223-8161, Weekly, 18,000, 24 pages, 10 x 15, $.85/line.

Ready Letter, Naval Reserve Readiness Command Region Thirteen, Building 1, Naval Base Great Lakes, Great Lakes, Il 60088, (312) 688-3652, Monthly, 1,700, 8 pages, 8 x 10½.

*Command Post Newspaper, Scott AFB, 612 E. State Street, O'Fallon, IL 62269, (618) 632-3643, Weekly, 11,000, 38 pages, 5 col. x 14½, $.25/line.

*Chanute This Week, Chanute AFB, c/o Rantoul Press, Inc., 1332 Harmon Drive, 1332 Harmon Drive, Rantoul, IL 61866, (217) 892-9613, Weekly, 7,500, 24 pages, 5 col. x 15, $.18/line.

AFCS Frequency Management Digest, HQ AFCS/DOYF, Scott AFB, Il 62225, (618) 256-5486, Bimonthly, 172, 25 pages, 8 x 10½.

AFCS Intercom, Scott AFB, IL 62225, (618) 256-4396, Biweekly, 17,500, 20 pages, 11¼ x 16.

Air Weather Service Observer, HQ., Military Airlift Command, Scott AFB, IL 62225, (618) 256-5003, Monthly, 2,600, 8 pages.

The MAC Flyer, Scott AFB, IL 62225, (618) 256-3534, Monthly, 11,000, 32 pages, 8 x 10½.

Military Airlift Command Information Newsletter, Scott AFB, IL 62225, (618) 256-3527, Bimonthly, 600, 8 pages, 8 x 10½.

The Tactical Communicator, Scott AFB, IL 62225, (618) 256-1110, Quarterly, 500, 30 pages, 8 x 10½.

General Orders, Daughters of Union Veterans of the Civil War, 503 S. Walnut Street, Springfield, IL 62704, (217) 544-0616, Quarterly, 20,000.

General Orders, National Woman's Relief Corps, Auxiliary to the Grand Army of the Republic, 629 S. Seventh, Springfield, IL 62703, (217) 522-4373, Bimonthly, 34,000.

Historical Reporter, Daughters of Union Veterans of the Civil War, 503 S. Walnut Street, Springfield, IL 62704, (217) 544-0616, Quarterly, 20,000.

Indiana

Army Administrator Magazine, Bldg. 600, Fort Benjamin Harrison, IN 46216, (317) 542-3567, Bimonthly, 5,000, 42 pages, 10½ x 8.

*The Harrison Post, Bldg. 601, Public Affairs Office, Fort Benjamin Harrison, IN 46216, (317) 542-3141, Weekly, 10,000, 16 pages, 10 x 15, $.255/line, Adv. to Hometown Publications, 2 So. Main Street, Fortville, IN 46040.

Military Media Review, Defense Information School, Public Affairs Office, Fort Benjamin Harrison, IN 46226, (317) 542-2173, Quarterly, 6,000, 32 pages.

*Grissom Contact, Grissom AFB, 113 N. Meridian Street, Greentown, IN 46936, (317) 628-3309, Weekly, 3,000, 16 pages, Tabloid, $.23/line.

*American Legion Magazine, American Legion, 700 N. Pennsylvania Street, Indianapolis, IN 46206, (317) 635-8411, Monthly, 2,700,000, 60 pages, 7½ x 10, $7,934/page.

The Army Dollar, Public Affairs Office, U.S. Army Finance and Accounting Center, Indianapolis, IN 46249, (317) 542-2219, Weekly, 3,100, 8 pages, 11 x 14.

The Beamrider, Naval Avionics Facility, 6000 E. 21st Street, Indianapolis, IN 46218, (317) 353-3163, Weekly, 3,000, 4 pages, 8 x 11.

Forty and Eighter, Forty and Eight (American Legion), 777 N. Meridian Street, Indianapolis, IN 46204, Monthly, 75,000.

Indiana Combat Veteran, Department of Indiana Veterans of Foreign Wars, 1402 N. Shadeland, Indianapolis, IN 46219, (317) 357-8367, Monthly, 59,800, 8 pages, 10½ x 14.

National News, American Legion Auxiliary, 777 N. Meridian Street, Indianapolis, IN 46204, (317) 635-8411, Bimonthly, 1,000,000.

SALutations, Sons of the American Legion, 700 N. Pennsylvania Street, Indianapolis, IN 46206, Quarterly, 24,688.

Kansas

*Fort Leavenworth Lamp, Fort Leavenworth, KS 66027, (913) 684-4051, 5,000, 20 pages, Tabloid, $.173/line.

Military Review, Funston Hall, Fort Leavenworth, KS 66027, (913) 684-5642, Monthly, 29,000, 112 pages, 9 x 6.

*Fort Riley Post, 814 N. Washington, Junction City, KS 66441, (913) 762-5000, Weekly, 10,060, 12 pages, 7 col. x 21.5, $.32/line.

*Aerospace Historian, Dept. of History, Eisenhower Hall, Kansas State University, Manhattan, KS 66506, (913) 532-6733, Quarterly, 4,000, 80 pages, 8½ x 11, $575/page.

*Military Affairs, Dept. of History, Eisenhower Hall, Kansas State University, Manhattan, KS 66506, (913) 532-6733, Quarterly, 2,400, 68 pages, 8½ x 11, $175/page.

Jetster, Civilian Personnel Office, McConnell AFB, KS 67221, (316) 681-5540, Monthly, 1,200, 4 pages, 8 x 10½.

Midwest Reporter, 9th Marine Corps. District Headquarters 10000 W. 75th, Shawnee Mission, KS 66204, (816) 926-6385, Monthly, 9,000, 16 pages, 8 x 11½.

Dad Talk, American War Dads, 117 W. Sixth Street, Topeka, KS 66603, (913) 354-8162, Semimonthly, 5,000.

*Contrails, 381 SMW/OI, McConnell AFB, KS 67221, (316) 681-5781, Weekly, 4,500, Tabloid, $.29/line, Adv. to Brazill Publications, 2318 East Central, Wichita, KS 67214.

Kentucky

DAV Magazine, Disabled American Veterans, 3725 Alexandria Pike, Cold Spring, KY 41076, (606) 441-7300, Monthly, 545,000.

DAV Auxiliary Magazine, Disabled American Veterans, 3725 Alexandria Pike, Cold Spring, KY 41076, (606) 441-7300, Monthly, 60,000.

*Fort Campbell Courier, Public Affairs Office, Fort Campbell, KY 42223 (502) 798-4017, Weekly, 15,050, 24 pages, 8 col. x 21½, $.15/line, Adv. to Box 829, Clarksville, KY 37040.

Armor—The Magazine of Mobile Warfare, U.S. Army Armor School, Box 0, Fort Knox, KY 40121, (502) 624-2249, Bimonthly, 18,300, 64 pages, 8¼ x 11.

*Inside the Turret, Public Affairs Office, Fort Knox, KY 40121, (502) 624-5341, Weekly, 22,000, 32 pages, 8 col. x 21¼, $.21/line.

The Beacon, Lexington-Blue Grass Army Depot, Lexington, KY 40511, (606) 293-3039, Bimonthly, 3,000, 4 pages, 15 x11½.

P.S. Magazine, Lexington, KY 40511, (606) 293-3367, Monthly, 130,000, 68 pages, 5 x 7.

Louisiana

*The Observer, Editor, 2nd Bomb Wing/OI, Barksdale AFB, LA 71110, (318) 456-3709, Weekly, 7,000, 24 pages, 11½ x 17½, $.334/line, Adv. to Bossier Newspaper Publishing Co., 409 Barksdale Blvd., Bossier City, LA 71111.

*Kisatchie Guardian, Public Affairs Office, Fort Polk, LA 71459, (318) 537-4033, Weekly, 10,000, 12 pages, Standard size, $.26/line.

Continental Marine, 4400 Dauphine Street, 4th Marine Division, New Orleans, LA 70146, (504) 948-1547, Monthly, 116,500, 8 pages, 10¼ x 14.

Digest, 4th Marine Aircraft Wing, 4400 Dauphine Street, New Orleans, LA 70146, (504) 948-1264, Monthly, 4,000, 8 pages, 8 x 10.

Naval Reservist News, Chief of Naval Reserve, 4400 Dauphine Street, New Orleans, LA 70146, (504) 948-1241, Monthly, 220,000, 8 pages, Tabloid.

Nolair News, Public Affairs Office, Naval Air Station, New Orleans, LA 70146, (504) 393-3260, Monthly, 2,500, 12 pages, 8½ x 11.

Safety Sentinel, 4th MAW/MARTC, 4400 Dauphine Street, New Orleans, LA 70146, (504) 948-1237, Bimonthly, 1,250, 16 pages, 8 x 10.

*Tiger Talk, 23 TFW/OI, England AFB, LA 71301, (318) 448-2401, Weekly, 7,000, 16 pages, 11½ x 15, $.25/line, Adv. to Tiger Talk, Box 973, Pineville, LA 71360, (318) 445-0148.

Bulletin, American Ex-Prisoners of War, c/o D.C. Wimberly, 711 Tenth, N.W., Springhill, LA 71075, Monthly.

Maine

*The Patroller, Naval Air Station, Brunswick, ME 04081, (207) 921-2327, Weekly, 4,000, 16 pages, Tabloid, $.17/line, Adv. to Coastal Journal, Down East Publications, 811 Middle Street, Bath, ME 04350.

Cutler Sparks, East Machias, ME 04630, (207) 259-8551, Bimonthly, 20 pages, 8 x 10½.

*The Hansconian, Hanscom AFB, ME 01731, (617) 861-5318, Weekly, 3,600, 8 pages, Tabloid, Adv. to Minute-Man Publications, Inc., 9 Meriam Street, Box 160, Lexington, ME 02173.

*Seacoast Flyer, 1 High Street, Kennebunk, ME 04043, (207) 985-2961, Weekly, 4,600, 20 pages, 16½ x 11¼, $.22/line.

*The Limelite, 42BMW/OI, Loring AFB., ME 04751, (207) 999-2170, Weekly, 5,000, 12 pages, $.187/line, Adv. to Northeast Publishers, Presque Isle, ME 04769.

News, AMVETS Auxiliary, Saco Road, Old Orchard Beach, ME 04064, (207) 934-4454, Quarterly, 25,000.

Acadian, Winter Harbor, ME 04693, (207) 963-5534, Biweekly, 650, 26 pages, 8 x 10½.

Maryland

*APG News, Harford Press Inc., 4 South Parke St., Aberdeen, MD 21001, (301) 272-2600, Weekly, 12,500, Standard, $.28/line, Adv. to Harford Press Inc., 4 South Parke St., Aberdeen, MD 21001.

Mini-Gram, Harry Diamond Laboratories, Department of the Army, 2800 Powder Mill Road, Adelphi, MD 20783, (202) 394-2654, Monthly, 1,500, 6 pages, 8 x 10.

*Capital Flyer, Andrews AFB, MD 20331, (301) 981-3606, Weekly, 12,000, 24 pages, 10 x 13½, $.30/line.

Systems Management Newsletter, HQ AFSC/SDDS, Andrews AFB, MD 20334, (301) 981-4373, Quarterly, 12 pages, 8½ x 10.

*Castle, 213 West Street, Annapolis, MD 21401, (301) 268-5000, Weekly, 15,000, Tabloid, $.50/line, Adv. to Capital Gazette Press Inc., 213 West Street, Annapolis, MD 21401.

*The Log, U.S. Naval Academy, Annapolis, MD 21412, Monthly, 5,500, 40 pages, 8 x 11, $120/page.

Naval Station Newsletter, U.S. Naval Station, Annapolis, MD 21402, (301) 267-2385, Biweekly, 1,100, 4 pages, 8 x 10.

*Shipmate, U.S. Naval Academy Alumni Association, Alumni House, Annapolis, MD 21402, (301) 263-4448, Monthly, 22,000, 96 pages, 8½ x 11, $475/page, Adv. to Gene Kramer, Box 96, Short Hills, NJ 07078.

*Soundoff, Ft. Meade, Capital-Gazette Newspapers, Inc., 213 West Street, Annapolis, MD 21401, (301) 269-0896, Weekly, 15,000, 14 pages, Standard, $.50/line.

*U.S. Naval Institute Proceedings, Annapolis, MD 21402, (301) 268-6110, Monthly, 65,334, 112 pages, 8⅜ x 11¼, Adv. to Donovan/Burke, Inc., 225 Park Avenue, New York, NY 10017.

Newsletter, Ladies Auxiliary, Military Order of the Purple Heart, United States of America, Harbor View Apts., 610 Light Street, Apt. 915, Baltimore, MD 21230, (301) 685-5604, 10 times/year.

The Purple Heart Magazine, Ladies Auxiliary, Military Order of the Purple Heart, United States of America, Harbor View Apts., 610 Light Street, Apt. 915, Baltimore, MD 21230, (301) 685-5604, Bimonthly, 3,000.

The Centerline, David W. Taylor Naval Ship R & D Center, Bethesda, MD 20084, (202) 227-2892, Biweekly, 4,200, 8 pages, 8½ x 11.

The Write On, USA Intelligence and Security Command Support Group, Fort Meade, MD 20755, (301) 677-7017, Monthly, 1,000, 8 pages, 8 x 10.

Order Wire, Department of the Army, Headquarters 7th Signal Command, Fort Ritchie, MD 21719, (301) 878-5874, Monthly, 2,300, 16 pages, 8½ x 11.

Ritchie News, Department of the Army, Headquarters 7th Signal Command, Fort Ritchie, MD 21719, (301) 878-5874, 8 pages, 11½ x 15.

Profile, Naval Ordnance Station, Indian Head, MD 20640, (301) 743-4202, Monthly, 2,700, 12 pages, 15 x 11⅜.

*Military Medicine, Box 104, Kensington, MD 20795, (301) 933-2801, Monthly, 10,000, 72 pages, 8½ x 11, $445/page, Adv. to Assn. of Military Surgeons of the U.S., Publisher, Box 104, Kensington, MD 20795.

Outlook, NAS, Patuxant River, MD 20670, (301) 863-3607, Quarterly, 4 pages, 8 x 11.

Tester, Naval Air Station, Patuxent River, MD 20670, Weekly, 6,500, 8 pages, 11½ x 16.

The Oak Leaf, Naval Surface Weapons Center, White Oak, Silver Springs, MD 20910, (202) 394-1810, Biweekly, 4,500, 6 pages, 11 x 15.

*Sea Power, 12005 Chesterton Drive, Upper Marlboro, MD 20870, (301) 249-7158, Monthly, 49,000, 64 pages, 8½ x 11, $1,200/page.

Massachusetts

*The Fort Devens Dispatch, 83 Fitchburg Road, Ayer, MA 01432, (617) 772-0777, Weekly, 4,000, 28 pages, 14 x 17, $.17/line.

Michigan

Legion News, 55 Victor, Highland Park, MI 48203, (313) 869-1090, Monthly, 5,000, 8 pages, $5/col. in.

The Motor Mouth, Tremonti Bldg., 426 Clinton Street, Detroit, MI 48226, (313) 226-7793, Monthly, 350, 8 pages, 8 x 10.

Smoke Signal, NAF Detroit, Mt. Clemens, MI 48043, (313) 466-5550, Monthly, 2,000, 4 pages, Tabloid.

*Northern Light, Box 429, Marquette, MI 49855, (906) 228-8920, Weekly, 4,200, Standard, $.171/line, Adv. to Action Shopper, Box 429, Marquette, MI 49855.

*Wurtsmith News, Wurtsmith AFB, MI 48753, (517) 739-6058, Weekly, 3,000, 12 pages, 11½ x 17½, $.121/line, Adv. to Richard McKnight, Box 271, Osoda, MI 48750.

Minnesota

Gopher Oversea'r, VFW HQ, Veterans Service Bldg., St. Paul, MN 55155, (612) 291-1757, 10 times/year, 80,000, 8 pages, 15 x 5 cols.

*The Minnesota Legionnaire, Veterans Service Bldg., St. Paul, MN 55155, (612) 291-1800, Biweekly, 125,000, 8 pages, 11½ x 16, $7.28/col. in.

Mississippi

Bulletin, Naval Oceanographic Office, NSTL Station, Bay St. Louis, MS 39522, (601) 688-4015, Monthly, 1,700, 12 pages, 8½ x 11.

*Blue Print, 516 Main Street, Columbus, MS 39701, (601) 328,2429, Weekly, 3,500, 12 pages, 9 x 13½, $.14/line, Adv. to Box 441, Columbus, MS 39701.

*The Seabee Courier, Naval Construction Battalion Center, Gulfport, MS 39501, (601) 865-2393, Biweekly, 10,000, 12 pages, Tabloid, $.25/line, Adv. to Gulfport Star Journal, Box 884, Gulfport, MS 39501.

Report of Adjutant-in-Chief, Sons of Confederate Veterans, Southern Station, Box 5164, Hattiesburg, MS 39401, (601) 545-1159, Monthly, 4,600.

*Keesler News, Keesler, AFB, MS 39534, (601) 377-4130, Weekly, 12,300, 28 pages, 14 x 10½, $.32/line, Adv. to Gulf Publishing Co., Box 4567, Biloxi, MS 39531.

*The Skyline, Box 23, Public Affairs Office, NAS Meridian, MS 39301, (601) 679-2318, Semimonthly, 2,500, 12 pages, 14 x 22, $.12 line, Adv. to Jack Bynum, The Meridian Star, 814 22nd Avenue, Meridian, MS 39301.

SupShip Navy Bulletin, Supervisor of Shipbuilding, Conversion and Repair, USN, Pascagoula, MS 39567, (601) 769-0186, Monthly, 5 pages, 8 x 10½.

Journal, Military Railway Service Veterans, c/o Mrs. Marie L. Godley, P.O. Box 529, Pass Christian, MS 39571, Quarterly, 5,000.

Missouri

Show Me Missouri Legionnaire, Box 179, Jefferson City, MO 65102, (314) 625-7979, 10 times/year, 60,000, 8 pages, 5 x 11½ pica cols.

Auxiliare, American War Dads Auxiliary, 1123 Scarritt Bldg., Kansas City, MO 64106, (816) 842-4210, Bimonthly, 2,396.

National Auxiliary Magazine, Ladies Auxiliary to the Veterans of Foreign Wars of the United States, 406 W. 34th Street, Kansas City, MO 64111, (816) 561-8655, 580,000.

National Organization Bulletin, United States Army Mothers, 818 Grand Avenue, Kansas City, MO 64106, (816) 221-4753, Monthly, 8,000.

*V.F.W. Magazine, Veterans of Foreign Wars of the U.S., Broadway at 34th Street, Kansas City, MO 64111, (816) 756-3390, Monthly, 1,812,000, 60 pages, 8½ x 10, $6,270/page.

Washington Action Reporter, Veterans of Foreign Wars of the U.S.A., V.F.W. Bldg., Kansas City, MO 64111, (816) 756-3390, Monthly.

Intercom, Stop 100 G, Richards-Gebaur AFB, MO 64030, (816) 348-3434, Biweekly, 21,500, 24 pages, 11 x 16½.

Richards-Gebaur Communicator, Richards-Gebaur AFB, MO 64030, (816) 348-2141, Weekly, 3,100, 8 pages, Tabloid.

Bulletin, Women's Auxiliary to the Military Order of the Cootie, c/o Yuba L. Clodfelter, 4521 Llama Lane, St. Joseph, MO 64506, (816) 233-5405, Monthly, 12,024.

General Orders, Women's Auxiliary to the Military Order of the Cootie, c/o Yuba L. Clodfelter, 4521 Llama Lane, St. Joseph, MO 64506, (816) 233-5405, Monthly, 12,024.

Gateway Reporter, Department of the Army, HQ, U.S. Army Troop Support & Aviation Materiel Readiness Command, 4300 Goodfellow Blvd., St. Louis, MO 63120, Bimonthly, 8,000, 12 pages, 11½ x 14½.

St. Louis Quality Assurance Newsletter, Defense Supply Agency, 1136 Washington Avenue, St. Louis, MO 63101, (314) 268-3695, Bimonthly, 500, 5 pages, 8 x 10½.

*Pulaski County Democrat, Box 578, Waynesville, MO 65583, (314) 774-6632, Weekly, 7,200, Standard, $.123/line.

*The Monitor, Whiteman AFB, MO 65301, Weekly, 3,600, 20 pages, 11⅜ x 15, $.133/line, Adv. to Sedalia Democrat Company, 700 S. Mass., Box 848, Sedalia, MO 65301.

Montana

Gemini, Malmstrom AFB, MT 59402, (406) 731-3091, Monthly, 1,500, 16 pages, 8 x 10½.

*Minuteman, Information Office, Malmstrom AFB, MT 59402, (406) 731-2408, Weekly, 5,000, 12 pages, 11½ x 16, $.18/line, Adv. to W.H. Chase, Publisher, 1108 4th Avenue N., Great Falls, MT 59401.

Nebraska

*The Nebraska Legionnaire, Box 5205, Lincoln, NE 68505, (402) 464-6338, Monthly, 60,000, 8 pages, 11½ x 16, $400/page.

Family Services Mirror, Offutt AFB, NE 68113, (402) 294-3111, Bimonthly, 100, 12 pages, 8 x 11½.

SAC Personnel Brief, Offutt AFB, NE 68113, (402) 294-3891, Bimonthly, 500, 20 pages, 8 x 10½.

*Air Pulse, Offutt AFB, 4875 F Street, Omaha, NE 68117, (402) 733-7300, Weekly, 36 pages, Tabloid, $.63/line.

*Offutt & SAC Base Guide and Directory, 4875 F Street, Omaha, NE 68117, (402) 733-7300, Annual, 8,000, 130 pages, 8¼ x 10¾, $650/page.

Nevada

*Desert Round Up, Box 828, Fallon, NV 89426, (702) 423-5445, Weekly, 1,000, Tabloid, $.18/line, Adv. to Lahontan Valley News, Box 828, Fallon, NV 89426.

The Rocket, Hawthorne Army Ammunition Plant, Hawthorne, NV 89415, (702) 945-7004, Biweekly, 1,400, 8 pages, 11¾ x 15.

*Bullseye, Office of Information, Nellis AFB, NV 89191, (702) 642-2567, Weekly, 5,000, 28 pages, 11 x 16, $.225/line, Adv. to North Las Vegas Valley Times, 1007 Cheyenne Avenue, North Las Vegas, NV 89030.

New Hampshire

*Seacoast Flyer, Pease AFB, NH 03801, (603) 436-1011, Weekly, 4,000, 20 pages, Tabloid, $.17/line, Adv. to Star Press, Inc., Kennebunk, ME 04043.

New Jersey

Eastern Arrow, MTMC Military Ocean Terminal, Bayonne, NJ 07002, (201) 858-6351, Monthly, 2,000, 8 pages, 8½ x 11.

*Arradcom Voice, US Army Armament Research & Development Command, Dover, NJ 07801, (201) 328-4008, Monthly, 9,000, 28 pages, $.343/line, Adv. to Rimol Associates, 92 Broadway, Suite 101, Denville, NJ 07834.

Air Scoop, Bldg. 26, PAO, Naval Air Engineering Center, Lakehurst, NJ 08733, (201) 323-2620, Weekly, 4,500, 4 pages, Tabloid.

The Banner, Sons of Union Veterans of the Civil War, Box 6193, Lawrenceville, NJ 08648, (609) 896-1948, Quarterly, 2,500.

*The Monmouth Message, HQ, US Army Electronics Command, Attn: DRSEL-IO, Fort Monmouth, NJ 07703, (201) 532-1584, Weekly, 12,000, 16 pages, $.27/line, Adv. to Monica Rafkin, One Register Plaza, Shrewsbury, NJ 07701.

*Fort Dix Post, 17 High Street, Mount Holly, NJ 08060, (609) 267-0285, Weekly, 10,500, 14 pages, $.32/line.

*McGuire Airtides, McGuire AFB, Fort Dix Road, Box 307, Pemberton, NJ 08068, (609) 894-2282, Weekly, 10,600, 32 pages, 5 x 15, $.16/line.

New Mexico

*Sunburst, Holloman AFB, Box 870, Alamorgordo, NM 88310, (505) 437-7120, Weekly, 6,000, Standard, $.17/line, Adv. to Alamogordo News, Box 870, Alamogordo, NM 88310.

*Focus, Box 25526, Albuquerque, NM 87125, (505) 843-7537, Weekly, 12,000, Tabloid, $.30/line, Adv. to Modern Press, Box 25526, Albuquerque, NM 87125.

*Mach Meter, 27 TFW/OI, Cannon AFB, NM 88101, (505) 784-3311, Weekly, 4,200, 24 pages, 14 x 9, $.17/line, Adv. to News, Portales, NM 88130.

*White Sands Missile Ranger, PAO, White Sands Missile Range, NM 88002, (915) 678-1134, Weekly, 12,000, 8 pages, 8 cols. x 21, $.17/line, Adv. to Sandy White, 109 Ridgecrest Drive, Las Cruces, NM 88001.

New York

Casemate, Public Affairs Office, Fort Hamilton, Brooklyn, NY 11252, (212) 836-4100, Biweekly, 2,000, 8 pages, 10 x 14.

Navy Resale System News Digest Magazine, Public Affairs, Navy Resale System Office, 3rd Avenue and 29th Street, Brooklyn, NY 11232, (212) 965-5237, Monthly, 27,000, 12 pages, 8 x 10½.

NSA Newsletter, Naval Support Activity Newsletter, 136 Flushing Avenue, Brooklyn, NY 11251, (212) 834-2794, Monthly, 350, 14 pages, 8 x 10½.

Anchor, American Cadet Alliance, 88-04 Little Neck Parkway, Floral Park, NY 11001, (212) 343-2943, 9 times/year, 2,000.

Retreads Blowout, Retreads, 40-07 154 Street, Flushing, NY 11354, (212) 359-0685, Semiannual, 1,500.

Roster of Officers, Retreads, 40-07 154th Street, Flushing, NY 11354 (212) 359-0685, Annual, 1,500.

The Standard Bearer, HQ, 1st Marine Corps District, 605 Stewart Avenue, Garden City, NY 11530, (516) 741-8000, Monthly, 4 pages, Tabloid.

The Defender, 21st Office of Information, Hancock Field, NY 13225, (315) 458-5500, Monthly, 1,500, 16 pages, 8½ x 11.

*Mohawk Flyer, Main Street, Holland Patent, NY 11354, (315) 865-4100, Weekly, 11,000, 20 pages, 6 col. x 16, $.13/line, Adv. to Steffen Publishing Co., Holland Patent, NY 11354.

QUAN, American Defenders of Bataan and Corregidor, c/o Alex L. Benishake, 34 Mountain View Avenue, Hurley, NY 12443, (914) 338-5268, Bimonthly, 1,750.

*Club and Food Service, Box 788, Lynbrook, NY 11563, (516) 887-1800, Monthly, 10,517, $600/page.

*Exchange and Commissary News, Box 788, Lynbrook, NY 11563, (516) 887-1800, Monthly, 11,502, $680/page.

*Eagle & Swan, 515 Madison Ave., New York, NY 10022, (212) 593-2223, Quarterly, 100,000, 48 pages, 7 x 10, $2,900/page.

*Army-Navy Store and Outdoor Merchandiser, 225 W. 34th Street, New York, NY 10016, (212) 279-0800, 3,550, Monthly, $590/page.

*Family, 275 Madison Avenue, New York, NY 10016, (212) 889-1200, 229,750, 10 times/year, 8¼ x 10⅞, $6,500/page.

Note Book, National Yeomen F., Ten Stuyvesant Oval, New York, NY 10009, (212) 475-4896, Quarterly, 1,300.

*Champlaner, 380th Bomb Wing, Office of Information, Plattsburgh, AFB, NY 12903, (315) 565-7072, Weekly, 3,200, 24 pages, Tabloid, $.28/line, Adv. to Press-Republican, 170 Margaret Street, Plattsburgh, NY 12901.

Annual Pictorial Magazine, Flying Tigers of the 14th Air Force Association, Box 285, Selden, NY 11784, (516) 698-1782, Bimonthly, 2,100.

Jing Bao Journal, Flying Tigers of the 14th Air Force Association, Box 285, Selden, NY 11784, (516) 698-1782, Bimonthly, 2,100.

News-Letter, American Legion Press Association, 152 Woodgate Road, Tonawanda, NY 14510, (716) 836-2035, Bimonthly, 1,200.

Depot Dispatch, Taughannock Publications, Box N, Trumansburg, NY 14886, (607) 387-9590, Monthly, 1,000, 8 pages.

The Salvo, Editor, US Army, Watervliet Arsenal, Watervliet, NY 12189, (518) 266-5418, Monthly, 3,000.

Admissions Office Information Bulletin, Director of Admissions, U.S. Military Academy, West Point, NY 10996, (914) 938-3423, 5 times/year, 2,300, 12 pages, 8 x 10¼.

*Pointer Magazine, West Point Military Reservation, West Point, NY 10996, Weekly, 6,000, $75/¼ page, Adv. to West Point Military Reservation, West Point, NY 10996.

North Carolina

Camp Lejeune Globe, Box 8438, Joint Public Affairs, Camp Lejeune, NC 28542, (919) 451-5655, 18,000, 12 pages, 14 x 12.

Camp Lejeune Trader, Base Special Services, Marine Corps Base, Camp Lejeune, NC 28542, (919) 451-2108, Weekly, 11,000, 3 pages, 8 x 10½.

Civilian Guidepost, Marine Corps Base, Camp Lejeune, NC 29542, (919) 451-1458, Biweekly, 3,100, 4 pages, 8 x 10.

*Golden Griffon, PAO, 108th Division (Training), US Army Reserves, Box 9307, Charlotte, NC 28299, (704) 333-8654, Quarterly, 3,500, 16 pages, 10 x 16, Adv. to Fulghum & Associates, Box 25828, Charlotte, NC 28212.

*Cherry Point Windsock, The Windsock, Marine Corps Air Station, Cherry Point, NC 28533, (919) 466-3547, Weekly, 12,500, 40 pages, $.23/line, Adv. to The Windsock, c/o Havelock Progress, Havelock, NC 28532.

*The Paraglide, Box 35395, Fayetteville, NC 28303, (919) 484-8153, Weekly, 21,000, Standard, $.36/line.

The Blue Falcon Monthly, Commander, 3d Bn (Abn) 325th, Inf., 82d Abn Div., Fort Bragg, NC 28307, (919) 396-6204, Monthly, 775, 15 pages, 8 x 13.

Persuader, HQ, 4th Psychological Operations Group, Fort Bragg, NC, 28307, (919) 396-8576, Monthly, 1,000, 28 pages, 8 x 10.

*Pointer, Box B, Jacksonville, NC 28540, (919) 353-1171, Monthly, 1,100, Tabloid, $.12/line, Adv. to Jacksonville Daily News, Box B, Jacksonville, NC 28540.

*Rotovue, Public Affairs Office, MCAS (H) New River, Jacksonville, NC 28545, Weekly, 5,000, 12 pages, 11½ x 16, $.12/line, Adv. to Bethel Printing, 1214 Gum Branch Road, Jacksonville, NC 28545.

Bulletin, Merrill's Marauders Association, c/o David Hurwitt, 8307 Cricket Lake Lane, Matthews, NC 28105, (704) 364-4819, 3 or 4 times/year, 800.

Hercules Herald, 317 TAW/OI, Pope AFB, NC 27281, (919) 394-4183, Biweekly, 1,500, 8 pages.

*Scope, Seymour Johnson AFB, NC 27531, (919) 736-5411, Weekly, 5,000, 20 pages, 11½ x 14½, $.173/line, Adv. to News Argus, Attn: Scope, Box 1718, Goldsboro, NC 27530.

North Dakota

*Leader, Box 475, Finley, ND 58230, (701) 524-1640, Weekly, 6,325, Tabloid, $.21/line.

*Sentry, Base Information Office, Minot AFB, ND 58701, (701) 463-2201, Weekly, 6,850, 28 pages, Tabloid, $.25/line, Adv. to BHG, Inc., Box 309, Garrison, ND 58540.

Ohio

Bugle Call, Ladies of the Grand Army of the Republic, 846 Main Street, Bridgeport OH 43912, (614) 635-1421, Bimonthly, 3,600.

Journal, Ladies of the Grand Army of the Republic, 846 Main Street, Bridgeport, OH 43912, (614) 635-1421, Annual 3,600.

DAV Magazine, Box 14301, Cincinnati, OH 45214, (606) 441-7300, Monthly, 580,000, 32 pages, 8¼ x 10¾.

DCSC Supply Line, Public Affairs Office DCSC-B, 3990 E. Broad St., Columbus, OH 43213, (614) 236-2328, Bimonthly, 5,000, 4 pages, 11¼ x 17½.

DESC Customer Information Bulletin, Defense Electronics Supply Center, Dayton, OH 45444, (513) 296-6421, Quarterly, 12,000, 4 pages, 8½ x 10.

Golden Sphinx, National Counter Intelligence Corps Association, 3969 Applweood Lane, Dayton, OH 45429, Quarterly, 3,500.

*The Skywrighter, Box 572, Dayton, OH 45409, (513) 294-0022, Weekly, 29,000, 40 pages, 10 x 16½, $3.70/col. in. (Wright-Patterson AFB)

Cootie Courier, Supreme Pup Tent, Military Order of the Cootie, Box 627, Fostoria, OH 44830, (419) 435-3150, Monthly, 31,000.

Gazette, General Society of Colonial Wars, 840 Woodbine Avenue, Glendale, OH 45246, (513) 771-8253, Semiannual, 4,200.

*On Target, Newark AFB, Mid-Ohio Industrial Park, Heath, OH 43055, (614) 929-2010, Bimonthly, 3,000, 12 pages, $.289/line, Adv. to Ace Newspapers, Mid-Ohio Industrial Park, Heath, OH 43055.

Encampment, Army and Navy Union, U.S.A., 1391 Main Street, Box 537, Lakemore, OH 44250, (216) 733-3113, Annual, 5,000.

Newsletter, Army and Navy Union, U.S.A., 1391 Main Street, Box 537, Lakemore, OH 44250, (216) 733-3113, 5 times/year, 5,000.

*Rickenbacker AFB, Skyhawk, 301 ARW/Information Office, Rickenbacker AFB, OH 43217, (614) 492-3981, Weekly, 6,500, 12 pages, 10 x 14½, $.32/line, Adv. to Grove City Publishing Co., 69 Grant Street, Grove City, OH 43123.

Service Star Newsletter, National Service Star Legion, 3496 Central Avenue, Shandyside, OH 43947, Quarterly, 2,000.

Air Force Engineering & Services Quarterly, AFIT/DEJ, Wright-Patterson AFB, OH 45433, (513) 255-2156, Quarterly, 5,000, 40 pages, 8½ x 11.

Oklahoma

*Altus Galaxy, Box 578, Altus, OK 73521, (405) 482-1221, Weekly, 3,500, Tabloid, $.18/line.

*Airscoop, Box 2261, Enid, OK 73701, (405) 234-6403, Weekly, 1,600, 12 pages, Tabloid, $.195/line, Adv. to The Haymaker Press, Inc., Box 2261, Enid, OK 73701.

*The Cannoneer, Public Affairs Office, Fort Sill, OK 73503, (405) 351-4163, Weekly, 13,000, 16 pages, 15 x 23, $.37/line, Adv. to Lawton Publishing Co., 3rd & A, Lawton, OK 73501.

Field Artillery Journal, Box 3131, Fort Sill, OK 73503, (405) 351-5121, Bimonthly, 17,000, 64 pages, 8½ x 11.

*Tinker Take Off, 4515 N. Santa Fe, Oklahoma City, OK 73118, (405) 521-0911, Weekly, 25,000, 32 pages, 17 x 11½, $.34/line. (Tinker AFB)

Oregon

Kingsley Times, Office of Information, 827th Air Defense Group, Kingsley Field, OR 97601, (503) 882-4411, Monthly, 800, 16 pages, 8½ x 11.

Pearl Harbor-Gram, Pearl Harbor Survivors Association, 8920 S.E. Clay Street, Portland, OR 97216, (503) 253-1691, Quarterly, 7,023.

Pennsylvania

Parameters—Journal of the US Army War College, USAWC, Carlisle Barracks, PA 17013, (717) 245-4943, Quarterly, 6,500, 100 pages, 8½ x 11.

Bulletin, United States Army Ambulance Service Association, 195 Barnsley Avenue, Huntingdon Valley, PA 19006, (225) 947-6507, three times/year, 400.

The Mechanews, PAO (703), Navy Ships Parts Control Center, Mechanicsburg, PA 17055, (717) 766-8511, Weekly, 6,500, 4 pages, 11 x 17.

The Banner, Auxiliary to Sons of Union Veterans of the Civil War, 4724 N. Mascher Street, Philadelphia, PA 19120, (215) 739-5796, Quarterly, 5,123.

COG "I" Bulletin, Naval Publications and Forms Center, 5801 Tabor Avenue, Philadelphia, PA 19120, (215) 697-2660, Quarterly, 4,300, 4 pages, 8 x 10½.

Food News, 2800 S. 20th St. (Air Force Services Office), Philadelphia, PA 19101, (215) 271-3835, Quarterly, 1,200, 8 pages, 8½ x 11.

Navmiro Manufacturing Technology Bulletin, Navmiro, Philadelphia, PA 19112, (215) 755-3992, Bimonthly, 4 pages, 8 x 10½.

News-Letter, Military Order of Foreign Wars of the United States, Hill-Pysick-Keith House, 321 S. Fourth Street, Philadelphia, PA 19106, Monthly, 3,000.

*Recon, Box 14602, Philadelphia, PA 19134, Monthly, 1,700, $40/page, Adv. to Recon Publications, Box 14602, Philadelphia, PA 19134.

Regional Reflections, Education and Training, Navy Regional Medical Center, 17th and Pattison, Philadelphia, PA 19145, (215) 755-8261, Monthly, 1,000, 4 pages, 12 x 16.

Tobyhanna Reporter, Tobyhanna Army Depot, Tobyhanna, PA 18466, (717) 894-8301, Monthly, 4,300, 8 pages, 11 x 15.

NADC Reflector, Code 011, Naval Air Development Center, Warminster, PA 18974, (215) 441-3067, 4,000, 6 pages, 11 x 16.

The Intercom, Naval Air Station, Willow Grove, PA 19090, (215) 441-1203, Monthly, 5,000, 8 pages, 11 x 16.

Rhode Island

Naval War College Review, Naval War College, Newport, RI 02840, (401) 841-2236, Quarterly, 17,000, 150 pages, 6 x 9.

*Newport Navalog, Newport NB, 101 Malbone Road, Newport, RI 02840, (401) 849-3300, Biweekly, 7,300, 12 pages, 11½ x 15½, $.30/line, Adv. to Edward A. Sherman Publishing Co., 101 Malbone Road, Newport, RI 02840.

South Carolina

*Shaw AFB RECON, Carolina Printing Center, Box 898, Cayce, SC 29033, (803) 796-1118, Weekly, 5,300, 20 pages, 5 col. x 14, $.25/line, Adv. to Carolina Printing Center, Box 898, Cayce, SC 29033.

*Bow Hook, Naval Station, Charleston, SC 29408, (803) 743-5379, Biweekly, 12,000, 20 pages, 12 x 15, $.23/line, Adv. to Island Publications, James Island, SC 29412.

The Brigadier, 3 Broad Street, Cadets of the Citadel, Charleston, SC 29401, Semimonthly, 1,900.

*Charleston Airlift Dispatch, 437th Military Airlift Wing, Charleston AFB, SC 29404, (803) 554-2453, Weekly, 6,000, 12 pages, 16 x 11, $.23/line, Adv. to Airlift Dispatch, Box 7505, North Charleston, SC 29406.

Money Matters, Navy Finance Office, Navy Base, Charleston, SC 29408, (803) 743-4100, Quarterly, 1,000, 6 pages, 8 x 10.

Newsgram, Naval Reserve Readiness Command Region Seven, Naval Base, Charleston, SC 29408, (803) 743-2631, Monthly, 700, 7 pages, 8 x 10½.

The Southerner, Facilities Engineering Command, 2144 Melbourne Street, Box 10068, Charleston, SC 29411, (803) 743-2723, Monthly, 700, 4 pages, 11 x 17.

The Trident, Naval Supply Center, Charleston, SC 29408, (803) 743-2972, Monthly, 2,500, 4 pages, 11 x 17.

*The Fort Jackson Leader, Command Information Office, Fort Jackson, SC 29207, (803) 751-5189, Weekly, 10,000, 24 pages, Tabloid, $.26/line, Adv. to Bruner Publishing Co., Box 98, Lexington, SC 29072.

*The Strand Sentry, 354 TFW/OI, Myrtle Beach AFB, SC 29577, (803) 238-7552, Weekly, 3,500, 20 pages, 6 x 14, $.19/line, Adv. to Sun News, Box 406, Myrtle Beach, SC 29577.

The Boot, PAO, MCRD Parris Island, SC 29905, (803) 525-3276, Weekly, 7,500, 12 pages, 5 col. x 15.

South Dakota

*The Plainsman, 44 SMW, Office of Information, Ellsworth AFB, SD 57706, (605) 399-2795, Weekly, 5,000, 40 pages, 5 col. x 13½, $.202/line, Adv. to Penn Publishing Co., 405 E. Omaha Street, Rapid City, SD 57701.

South Dakota Legion News, Box 67, Watertown, SD 57201, (605) 886-3604, Monthly, 31,000, 8 pages, 11½ x 17.

Tennessee

P.T. Boat All Hands, P.T. Boats, Inc., Box 109, Memphis, TN 38101, (501) 735-4473, 3 times/year, 7,800.

Reserve Memphis News, N.A.R. P.A.O., N.A.S., Memphis, TN 38054, (901) 872-5643, Quarterly, 2,500, 15 pages, 8½ x 11.

*Bluejacket, Naval Air Station, Memphis, S-l, Millington, TN 38054, (901) 872-5529, Weekly, 10,000, 16 pages, Tabloid, $.26/line, Adv. to Bluejacket, 3750 N. Watkins St., Memphis, TN 38127.

The Navy Human Resource Journal, NAS, Memphis, Millington, TN 38054, (901) 872-5155, Quarterly, 20 pages, 8 x 11½.

Texas

*Texas Legion News, Box 789, Austin, TX 78767, (512) 472-4138, Bimonthly, 80,000, 16 pages, 8½ x 11, $600/page.

*The Beehive, PAO, NAS Chase Field, Beeville, TX 78103, (512) 354-2464, Weekly, 3,000, 4 pages, 9 x 14, $.18/line, Adv. to Beeville Publishing Co., Inc., Beeville, TX 78102.

*Bergstrom Air Force Base Guide, 67th TRW/OI, Bergstrom AFB, TX 78743, (512) 385-4100, Annual, 4,500, 40 pages, 8 x 11, Adv. to Military Publishers, Box 1000, San Diego, CA 92138.

*The Jet Gazette, 67 TRW/OI, Bergstrom AFB, TX 78743, (512) 385-4100, Weekly, 6,000, 20 pages, 10 x 14, $.16/line, Adv. to Bluebonnet Publications, 901 S. Hwy 281, Burnet, TX 78611.

Discovery, Office of Information, Brooks AFB, TX 78235, (512) 536-3234, Biweekly, 1,000, 8 pages, 11 x 15.

*Carswell Sentinel, OI, 7th Bomb Wing, Carswell AFB, TX 76127, (817) 738-3511, Weekly, 6,500, 24 pages, 10¼ x 13½, $.249/line, Adv. to All Church Press, Inc., Box 1159, Fort Worth, TX 76101.

The Aircraftsman, Corpus Christi Army Depot, SOSCC-N, Bldg. 8, Corpus Christi, TX 78419, (512) 939-3626, Biweekly, 3,500, 8 pages, 11½ x 15.

*The Beam, PAO, Bldg. 3, NAS Corpus Christi, TX 78419, (512) 939-2568, Biweekly, 4,500, 8 pages, 11 x 14, $.23/line, Adv. to Flour Bluff Publishing Company, 538-B Caribbean, Corpus Christi, TX 78418.

DCASR Dallas Quality Assurance Newsletter, 500 South Ervay Street, Dallas, TX 75201, (214) 670-9471, Quarterly, 600, 15 pages, 8 x 10½.

Sky Ranger, PAO, NAS, Dallas, TX 75211, (214) 266-6140, Monthly, 2,500, 8 pages, 8 x 10½.

*Border Eagle, Drawer C, Del Rio, TX 78840, (512) 775-1551, Weekly, 2,500, 12 pages, 8 x 21, $.08/line, Adv. to Del Rio News Herald, Drawer C, Del Rio, TX 78840.

*Dyess Peacemaker, Dyess AFB, TX 79607, (915) 696-4375, Weekly, 15,000, 24 pages, 11 x 14, Adv. to Feather Publishing, 858 N. 1st, Abilene, TX 79601.

Air Defense Magazine, US Army Air Defense School, Box 5600, Fort Bliss, TX 79916, (915) 568-5240, Quarterly, 12,000, 64 pages, 8½ x 11.

*Ft. Bliss Monitor, 3704 Hercules, El Paso, TX 79904, (915) 779-7518, Weekly, 20,000, Tabloid, $.252/line, Adv. to Ward Publishing Co., 3704 Hercules, El Paso, TX 79904.

The High Flyer, Commander, 1st ASA AVN CO. (R), Fort Bliss, TX 79916, (806) 568-8922, Monthly, 150, 20 pages, 8 x 10.

Sunburst, PAO, 13th COSCOM, Fort Hood, TX 76544, (817) 685-5812, Monthly, 1,300, 12 pages, 8 x 10½.

Dateline-TCATA, TRADOC Combined Arms Test Activity, PAO, W. Fort Hood, TX 76544, (817) 532-9725, Monthly, 500, 14 pages, 8½ x 11.

*Texas Veterans News, Box 14409, Fort Worth, TX 76117, (817) 924-3216, Bimonthly, 10,000, 24 pages, 6 col., $79/half page.

Full Deflection, American Fighter Aces Association, c/o William N. Hess, P.O. Box 61268, Houston, TX 77208, (713) 237-4756, Quarterly, 515.

*Flying K, PAO, NAS Kingsville, TX 78363, (512) 595-6146, Semimonthly, 2,500, 8 pages, Tabloid, $.40/line, Adv. to Jim Allen, Kingsville Publishing Co., 105 S. 5th St., Kingsville, TX 78363.

*The Roundup, Box 2415, Lubbock, TX 79408, (806) 763-4551, Weekly, 4,200, 16 pages, 11½ x 17½, $.20/line.

Air Force Officers Career Newsletter, Randolph AFB, TX 78148, (512) 652-3568, Bimonthly, 60,000, 4 pages, 8 x 10.

Manpower and Organization Newsletter, Randolph AFB, TX 78148, (512) 652-5916, Quarterly, 600, 16 pages, 8 x 10½.

USAF News for Retired Personnel, Randolph AFB, TX 78148, (512) 652-3517, Bimonthly, 458,000, 6 pages, 8 x 10½.

*Goodfellow Monitor, Goodfellow AFB, Box 1332, San Angelo, TX 76902, (915) 949-1422, Weekly, 3,000, 12 pages, 6 col. x 14, $.21/line.

Army-Navy-Air Force Courier, 105 N. Alamo, 818 Gibbs Bldg., San Antonio, TX 78205, Monthly.

Daedalus Flyer, Order of Daedalians, Bldg. 1635, Kelly AFB, San Antonio, TX 78241, (512) 924-9285, Quarterly, 9,200.

*Kelly Observer, Box 27040, 7137 Military Drive West, San Antonio, TX 78227, (512) 673-4457, Weekly, 36 pages, 10¼ x 16½, $.40/line, (Kelly AFB).

*Lackland Tailspinner, Lackland AFB, TX 78236, (512) 671-2907, Weekly, 36 pages, 10¼ x 16½, $.40/line, Adv. to Box 27040, San Antonio, TX 78227.

*Medical Patriot, Box 27040, 7137 Military Drive West, San Antonio, TX 78227, (512) 673-4457, Weekly, 36 pages, 10¼ x 16½, $.35/line.

NCOA Journal, Box 2268, San Antonio, TX 78298, (512) 658-8831, Monthly, 132,000, 20 pages, 11½ x 15½.

*Vanguard Magazine, Box 2268, San Antonio, TX 78298, Monthly, $1,575/page, Adv. to Assc. of the United States of America, Publisher, Box 2268, San Antonio, TX 78298.

Ranger Journal, Ranger Battalions Association World War II, 4609 NASA Road 1, Apt. 309, Seabrook, TX 77586, (713) 334-5678, Biannual, 1,250.

*The Sheppard Senator, 011 N/1, Sheppard AFB, TX 76311, (817) 851-2733, Weekly, 10,000, 10 pages, 8 col. x 21½, $.38/line, Adv. to Times Publishing Co., Box 120, Wichita Falls, TX 76307.

*Ft. Hood Sentinel, Box 868, Temple, TX 76501, (817) 778-4444, Weekly, 20,000, Standard, $.17/line, Adv. to Temple Telegram, Box 868, Temple, TX 76501.

*Red River Review, Red River Army Depot, Texarkana, TX 75501, (214) 838-3143, Monthly, 11,000, 16 pages, 12 x 15, $.189/line, Adv. to Mrs. Evelyn Belk, 915 Pine Street, Texarkana, TX 75501.

*Wingspread, Randolph AFB, 122 E. Byrd St., Universal City, TX 78148, (512) 658-7424, Weekly, 7,000, Tabloid, $.251/line, Adv. to Herald Publishing Co. of Texas, 122 E. Byrd, Universal City, TX 78148.

*Ft. Sam Houston News Leader, 122 E. Byrd, Universal City, TX 78148, (512) 658-7424, Weekly, 12,000, Adv. to Herald Publishing Co. of Texas, 122 E. Byrd, Universal City, TX 78148, $.35/line.

Utah

*Desert Sampler, US Army Dugway Proving Ground, Attn.: STEDP-PA, Dugway, UT 84022, (801) 522-2116, Bimonthly, 1,500, 8 pages, Standard, $.145/line, Adv. to Tooele Transcript Bulletin, 58 N. Main, Tooele, UT 84074.

*Hill Top Times, Hill AFB, UT 84406, (801) 777-7321, Weekly, 15,500, 32 pages, 11½ x 15, $.29/line, Adv. to Mor Media Sales, Inc., 1150 W. Riverdale Rd., Ogden, UT 84403.

*Hub, Box 423, Roy, UT 84067, (801) 394-9655, Biweekly, 5,500, Tabloid, $.12/line, Adv. to Box 423, Roy, UT 84067.

Vermont

*Military Journal, 218 Beech St., Bennington, VT 05201, (802) 447-0313, Bimonthly, 1700, 48 pages, 8½ x 11, $8/col. in.

Virginia

Defense Management Journal, OASD (MRA&L), Cameron Station, Alexandria, VA 22314, (202) 274-7558, Bimonthly, 30,000, 48 pages, 7⅞ x 10¼.

DFSC Fuel Line, Defense Fuel Supply Center, Cameron Station, Alexandria, VA 22314, (202) 274-6489, Quarterly, 5,000, 35 pages, 8 x 10½.

General Orders, Veterans of World War I of U.S.A., 916 Prince Street, Alexandria, VA 22314, (703) 836-3060, Monthly, 115,000.

*Sojourner, 4600 Duke Street, Alexandria, VA 22304, (703) 751-0453, Bimonthly, 9,000, $160/page, Adv. to National Sojourners Inc., 4600 Duke Street, Suite 300, Alexandria, VA 22304.

Soldiers, Alexandria, VA 22314, (202) 274-6672, Monthly, 206,000, 56 pages, 8½ x 11.

The Torch, Veterans of World War I of U.S.A., 916 Prince Street, Alexandria, VA 22314, (703) 836-3060, Monthly, 115,000.

Torch, National Ladies Auxiliary to Veterans of World War I of the U.S.A., 114 N. Alfred Street, Alexandria, VA 22314, (703) 548-3616, Monthly, 88,000.

American Forces Press Service Clipsheet, 1117 N. 19th Street, Arlington, VA 22209, (202) 694-8059, Weekly, 2,500, 3 pages, 11 x 16.

General Orders, Legion of Valor of the United States of America, 621 S. Taylor Street, Arlington, VA 22204, (703) 525-6637, 10 times/year, 900.

The Journal of the Army Intelligence and Security Command, HQ INSCOM, Arlington, VA 22212, (202) 222-5496, Monthly, 2,250, 12 pages, 8 x 10.

The Marine's Magazine, Marine Corps League Auxiliary, 933 N. Kenmore Street, Suite 315, Arlington, VA 22201, (703) 524-5026, Monthly, 3,500.

Naval Research Reviews, Office of Naval Research, Code 730, Arlington, VA 22217, (202) 696-4254, Monthly, 10,000, 32 pages, 6 x 11.

Navy Editor Service, Navy Internal Relations Activity, Room 1044, Crystal Plaza #6, 2221 Jefferson Davis Hgwy., Arlington, VA 20360, (202) 692-2564, Biweekly, 3 pages, 10 x 15.

*Navy News, Box 69, Arlington, VA 22210, (703) 524-8910, Weekly, 20,500, 24 pages, Tabloid, $.59/line.

*Pentagram News, Box 69, Arlington, VA 22210, (703) 524-8910, Weekly, 30,200, 32 pages, Tabloid, $.59/line.

The Purple Heart, Military Order of the Purple Heart of the U.S.A., Box 9286, Arlington, VA 22209, (703) 525-6637, Bimonthly, 12,000.

Recruiter, 4015 Wilson Blvd., Arlington, VA 22202, (202) 692-4200, Monthly, 10,000, 12 pages.

SAM, 1117 N. 19th St., Arlington, VA 22209, (202) 694-5070, Monthly, Tabloid, 236,000, 20 pages.

*Quantico Sentry, Triangle Plaza, Dunfries, VA 22026, (703) 875-8700, Weekly, 11,000, 32 pages, 8 cols. (14 in.) x 21½ in., $.28/line, Adv. to Potomac Newspapers, Inc., Triangle Plaza, Dunfries, VA 22026.

The Engineer, PAO, Fort Belvoir, VA 22060, (703) 664-2838, Quarterly, 8,000, 48 pages, 8¼ x 11.

USACSC Interface, USACSC HQ. PAO C-9, Fort Belvoir, VA 22060, (703) 664-5018, Bimonthly, 2,000, 32 pages, 8 x 10.

7th World, 7th Transportation Group (TML), Fort Eustis, VA 23604, (804) 878-3390, Monthly, 4,000, 24 pages, 8 x 10.

Army Logistician Magazine, U.S. Army Logistics Management Center, Fort Lee, VA 23801, (804) 734-4342, Bimonthly, 55,000, 48 pages, 8⅜ x 10⅞.

*The Flyer, OI/1st Tactical Fighter Wing, Langley AFB, VA 23665, (804) 764-2018, Weekly, 10,000, 16 pages, 11½ x 15, $.49/line, Adv. to Ms. Randi Fontaine, Program Service, Inc., 3 Koger Executive Center, Norfolk, VA 23502.

Langley Outlook, Langley AFB, VA 23665, (804) 764-2152, Monthly, 10 pages, 8 x 10½.

Emphasis and Trends, Southern Field Division, Office of Civilian Personnel, Bldg. A-67, Naval Station, Norfolk, VA 23511, (804) 444-2341, Bimonthly, 4 pages, 8 x 10½.

The Gator, Bldg. 3006, NAB Little Creek, Norfolk, VA 23521, (804) 464-7923, Biweekly, 12,000, 12 pages, 11½ x 15.

Navy Lifeline, Naval Safety Center, NAS, Norfolk, VA 23511, (804) 444, 1321, Bimonthly, 40,000, 24 pages, 8½ x 11.

*Soundings, Program Service, Inc., 3 Koger Executive Center, Suite 103, Norfolk, VA 23502, (804) 461-7663, Weekly, 22,000, 28 pages, 9¾ x 14, $.55/line.

Sar-Tar News, Naval Air Reserve Unit, Naval Air Station, Norfolk, VA 23511, (804) 444-2414, Monthly, 1,600, 4 pages, 8½ x 11½.

Supply Chest, Naval Supply Center, Norfolk, VA 23512, (804) 444-2585, Biweekly, 5,500, 8 pages, 8½ x 11.

*Wheel, 3 Koger Executive Center, Suite 103, Norfolk, VA 23502, (804) 461-7663, Weekly, 10,000, Tabloid, $.49/line, Adv. to Wheel, 3 Koger Executive Center, Suite 103, Norfolk, VA 23502.

*Leatherneck Magazine, Box 1775, Quantico, VA 22134, (703) 640-3171, Monthly, 64,661, 64 pages, 8½ x 11, $800/page.

*The Marine Corps Gazette, Box 1775, Quantico, VA 22134, (703) 640-6161, Monthly, 25,000, 96 pages, 8½ x 11, $625/page, Adv. to Advertising Dept., 1732 Eye Street, N.W., Washington, D.C. 20006.

Dockside Gosport, Code 103, PAO, Norfolk Naval Shipyard, Portsmouth, VA 23709, (804) 393-3592, Biweekly, 12,500, 15 pages, 8 x 11.

Navelex News, Box 55, Portsmouth, VA 23705, (804) 393-5594, Monthly, 500, 14 pages, 8 x 10½.

NAVSEACENLANT, Naval Sea Support Center, Atlantic, St. Juliens Creek Annex, Portsmouth, VA 23702, (804) 393-5323, Bimonthly, 600, 14 pages, 8 x 10½.

*Armed Services News, 2403 Virginia Beach Blvd., Virginia Beach, VA 23454, (804) 486-8000, Monthly, 35,000, 32 pages, Tabloid, $11.20/col. in., Adv. to Armed Services Newspaper Corporation, Navy News Bldg., 2403 Virginia Beach Blvd., Virginia Beach, VA 23454. *Note:* This newspaper group consists of four newspapers (Army Pictorial News, Navy Pictorial News, USAF Pictorial News and USMC Pictorial News), plus a supplement (Armed Services Pictorial News) which is the second section of all of them.

The Blue Ridger, HQ, 80th Div., 6700 Strathmore Rd., Richmond, VA 23224, (804) 229-1000, Semimonthly, 2,500, 24 pages, 8 x 10.

UDC Magazine, United Daughters of the Confederacy, 328 North Blvd., Richmond, VA 23220, (804) 355-1636, Monthly, 35,000.

The Jet Observer, Naval Air Station Oceana, Virginia Beach, VA 23460, (804) 425-3131, Bimonthly, 5,000, 8 pages, 11½ x 15.

Monthly Memo, Navy Wives Clubs of America, 4939 Casablanca Road, Virginia Beach, VA 23455, Monthly, 3,000.

Navy Wives News, Navy Wives Clubs of America, 4939 Casablanca Road, Virginia Beach, VA 23455, Monthly, 3,000.

*The Booster, Box 261, 224 Moorstown Road, Williamsburg, VA 23185, (804) 229-7200, Monthly, 3,500, 16 pages, $3.49/col. in.

Washington

Puget Soundings, Naval Supply Center, Puget Sound, Bremerton, WA 98314, (206) 478-7504, Monthly, 6 pages, 8½ x 11.

*Fairchild Times, 92 Bomb Wing/OI, Fairchild AFB, WA 99011, (509) 247-5704, Weekly, 6,000, 16 pages, 10 x 16, $.16/line, Adv. to Times Publishing Co., 92 Bomb Wing/OI, Fairchild AFB, WA 99011.

*Ranger, PAO, HQ 9th Inf. Div. and Fort Lewis, Fort Lewis, WA 98433, (206) 967-5738, Weekly, 20,000, 32 pages, $.31/line.

Keynotes, Code 064, Naval Torpedo Station, Keyport, WA 98345, (206) 396-2527, Semimonthly, 3,500, 8 pages, 8½ x 11.

*The Whidbey Crosswind, Naval Air Station, Whidbey Island, Oak Harbor, WA 98278, (206) 257-2316, Weekly, 5,500, 16 pages, 6 col. x 15, $.215/line, Adv. to Whidbey Press Publishing Co., Drawer F, Oak Harbor, WA 98277.

Ketch 22, Bldg. 9, NSA, Seattle, WA 98115, (206) 527-3338, Quarterly, 2,000, 8 pages, 11 x 16.

*Northwest Airlifter, McChord AFB, 9505 Bridgeport Way SW, Tacoma, WA 98499 (206) 582-2100, Weekly, 7,900, 24 pages, 10¼ x 13½, $.18/line.

West Virginia

*West Virginia Legionnaire, 2016 Kanawha Blvd. E., Charleston, WV 25311, (304) 343-7591, Monthly, 27,000, 8 pages, 11½ x 22, $810/page.

Legion of Guardsmen Magazine, Legion of Guardsmen Auxiliaries, 2610 Riverview Drive, Parkersburg, WV 26101, (304) 428-1142, Bimonthly, 1,000.

General Orders, Legion of Guardsmen Auxiliaries, 2610 Riverview Drive, Parkersburg, WV 26101, (304) 428-1142, Quarterly, 1,000.

Wisconsin

Badger Legionnaire, 812 E. State Street, Milwaukee, WI 53202, (414) 271-1940, Monthly, 93,000, 12 pages.

Sound Off, China-Burma-India Veterans Association, P.O. Box 1848, Milwaukee, WI 53201, (414) 677-2592, Quarterly, 2,500.

Wyoming

*Sentinel, Warren AFB, WY 82001, (301) 775-3610, Weekly, 10,000, 16 pages, 10 x 14, $.14/line, Adv. to Star Publications, 1613 Evans, Cheyenne, WY 82001.

Outside of the U.S.

Air Post Offices
(publications arranged alphabetically)

The Aegean Breeze, TUSLOG Det 118/OI, A.P.O. New York 09224, Weekly, 950, 6 pages, 8½ x 13.

The Anatolian Times, HQ TUSLOG/OI, A.P.O. New York 09254, Weekly, 1,800, 6 pages, 8 x 13.

The Antenna, Office of Information, 6950th Scty. Gp., A.P.O. New York 09193, Weekly, 1,200, 4 pages, 13 x 17½.

Arrow, Headquarters, 8th Infantry Division, A.P.O. New York 09111, Biweekly, 10,000, 8 pages.

The Artician, Commander, U.S. Army Cold Regions Test Center, Attn: STECR-TD-PAO, A.P.O. Seattle 98733, Biweekly, 200, 4 pages, 8 x 10½.

Berlin Observer, Public Affairs Office, USCOB, A.P.O. New York 09742, Weekly, 4,300, 8 pages, 13 x 26.1 cm.

The Blazer, HHC (S-3) 54th Engineer Battalion, A.P.O. New York 09026, 500, 10 pages, 10½ x 8½.

Bomb Blast, Editor, Miesan Army Depot, A.P.O. New York 09059, 550, 24 pages, 6½ x 8.

The Command Review, 19th Support Command, PAO, A.P.O. San Francisco 96212, Semimonthly, 3,500, 8 pages, 10½ x 15.

Community Life, HW, 1st Armored Division, West Germany, A.P.O. New York 09326, Monthly, 9,000, 4 pages, 11¼ x 15¼.

Crossroads, 1605th ABW/OI, A.P.O. New York 09406, Biweekly, 1,800, 8 pages, 11½ x 15.

Defender, 51st Composite Wing (Tactical)/OI, A.P.O. San Francisco 96570, Weekly, 3,000, 8 pages, Tabloid.

Donau Republic, HQ DET, USMCA, Neu Ulm, A.P.O. New York 09035, Biweekly, 3,000, 4 pages, 10½ x 15.

The Dragoon, Editor, 2d ACR, PAO, A.P.O. New York 09093, Monthly, 3,000, 4 pages, Tabloid.

EvrArmy Magazine, HQ, AFN-Europe, A.P.O. 09757, Monthly, 50,000, 32 pages, 7¾ x 10.

Far East Network Newsletter, 6204th Broadcasting Squadron (PACAF), A.P.O. San Francisco 96328, Quarterly, 140, 6 pages, 8 x 10½.

Five x Five, CDR, 32d Sig. Bn. (C), A.P.O. New York 09757, 750, 12 pages, 8 x 10.

The Frankfurt Chronicle, Box 14, Frankfurt Military Community, A.P.O. New York 09710, Weekly, 8,000, 8 pages, 4 col. x 14.

Frontline, PAO, 3d Infantry Division, A.P.O. New York 09036, 8 pages, 11 x 15.

Fuji Flyer, 475th ABW/OI, A.P.O. San Francisco 96328, Weekly, 5,000, 12 pages, 5 x 14.

Gauntlet, HQ, 38th ADA BDE, Attn.: PAO, A.P.O. San Francisco 96570, Semimonthly, 2,000, 8 pages, 11¼ x 15.

Ironsides, HQ, 1st Armored Division, A.P.O. New York 09326, Semimonthly, 28,000, 8 pages, 11¼ x 15¼.

Kadena Falcon, OI, 18th TFW, Kadena AB, Japan, A.P.O. San Francisco 96239, Weekly, 15,000, 12 pages.

Main Megaphone, Schweinfurt Military Community, A.P.O. New York 09033, Weekly, 3,000, 4 pages, 16 x 22.

The Maintainer, USAG-CH-PAO (Camp Humphreys), A.P.O. San Francisco 96271, 5,000, 8 pages, Tabloid.

Marmara Messenger, TUSLOG Det 94/OI, A.P.O. New York 09324, Weekly, 800, 4 pages, 11 x 16.

The Medcom Examiner, Hqs. U.S. Army Medical Command, Europe, A.P.O. New York 09403, Monthly, 10,000, 16 pages, 11 x 15.

The Misawan, NAVSECGRUACT, PAO (63), A.P.O. San Francisco 96519, Monthly, 5,000, 12 pages, 10½ x 7½.

Newsbeam, HQ, 11th AD Signal Battalion, Attn: AETL-SS-IO, A.P.O. New York 09175, Monthly, 600, 12 pages, 8 x 13.

Northern Light, 6920th Security Wing, Office of Information, A.P.O. San Francisco 96519, Weekly, 8 pages, Tabloid.

The Northern Scanner, 600 TCG/OI, A.P.O. New York 09669, Semimonthly, 300, 8 pages, 9½ x 13½.

The Oracle, A.P.O. New York 09223, Semimonthly, 2,000.

The Pershing Cable, 56 FA (West Germany), A.P.O. New York 09281, Monthly, 3,000, 8 pages, Tabloid.

Phantom Forum, 81 TFW/OI, A.P.O. New York 09755, Weekly, 3,500, 4 pages, 10 x 14.

Photogram, OI, 10th Tactical Reconnaissance Wing, A.P.O. New York 09238, Weekly, 2,000, 4 pages, 14 x 10.

Pillars and Posts, PAO, 3d Infantry Division, A.P.O. New York 09036, 6,000, 4 pages, 11 x 15.

Pirmasens Community Times, PAO, USMCA, A.P.O. New York 09189, Semimonthly, 2,500, 8 pages, 11¾ x 8¼.

*The Port Reporter, Public Affairs Office, USASPTGRP-NDL, A.P.O. New York 09069, Monthly, 2,500, 8 pages, Tabloid, $2/col. in., Adv. to H. Eversmann, Azeigen-Agentur Ditzen K.G., 285 Bremerhaven Hafenstr. 140.

Raider Magazine, 401 TFW/OI, A.P.O. New York 09283, Torrejon Air Base, Madrid, Spain, Biweekly, 4,000, 24 pages.

The Relay, USACC Taiwan, Box 3, A.P.O. San Francisco 96263, Monthly, 500, 6 pages, 10 x 14.

Rheinlander, Headquarters, 8th Infantry Division, Information Section, A.P.O. New York 09111, Bimonthly, 11,000, 8 pages, Tabloid.

SASCOM Courier, Headquarters 59th Ord. Bde. (PAO), A.P.O. New York 09189, Monthly, 2,500, 12 pages, 8 x 11.

Short Wave, 18 CSG/DPCE, A.P.O. San Francisco 96239, Monthly, 2,500, 12 pages, 8 x 5¼.

Soesterberg Interceptor, 32 TFS/OI, A.P.O. New York 09292, Biweekly, 800, 12 pages, 8 x 10.

The Spangled Banner, Spangdahlem A.B., 52 TFW/OI, West Germany, A.P.O. New York 09123, Weekly, 2,000, 6 pages.

Spearhead, HQ, 3rd Armored Division, A.P.O. New York 09039, Weekly, 10,500, 8 pages, Tabloid.

The Star, HQ, 95th MP BN, A.P.O. New York 09086, Monthly, 1,000, 14 pages, 8 x 10.

The Stars and Stripes, A.P.O. New York 09175, Daily, 110,000, 28 pages, Tabloid.

*The Traveller, 513th Tactical Airlift Wing, RAF Mildenhall, England, A.P.O. New York 09127, Weekly, 3,500, 12 pages, 11 x 16, Adv. to Saxon House, 1 Vince Road, Diss, Suffolk, England.

Tortt, HQ, U.S. Army Garrison Honshu, A.P.O. San Francisco 96343, Biweekly, 2,000, 12 pages, 11 x 14.

Typhoon Topics, 18 CSG/DPCER, A.P.O. San Francisco 96239, Monthly, 1,800, 8 pages, 5 x 8.

USAFE Civilian Personnel Letter, HQ/USAFE/DPC, A.P.O. New York 09012, Quarterly, 300, 6 pages, 8 x 10¼.

The Vigileer, Office of Information, 40th Tactical Group, A.P.O. New York 09293, Biweekly, 1,500, 8 pages, 26 cm. x 36 cm.

Zoeckler Zephyr, U.S. Army Field Station Korea, A.P.O. San Francisco 96271, Monthly, 1,500, 20 pages.

Fleet Post Offices (publications arranged alphabetically)

Adak Eagle's Call, Public Affairs Office, Box 2, F.P.O. Seattle 98791, Weekly, 2,700, 20 pages, 8 x 10.

Big E Magazine, Public Affairs Office, USS Enterprise, CVW-65, F.P.O. San Francisco 96601, Bimonthly, 5,000, 32 pages, 8 x 10.

El Tiburon, USNSGA Sabana Seca, F.P.O. New York 09555, Weekly, 500, 12 pages, 8 x 10.

European Scientific Notes, Office of Naval Research Branch Office, Box 39, F.P.O. New York 09510, Monthly, 7,000, 35 pages, 8 x 10½.

Focus on Four, Public Affairs Office, U.S. Navy Mobile Construction, Battalion Four, F.P.O. San Francisco, CA 96601, Monthly, 1,100, 34 pages.

Guantanamo Gazette, PAO, Box 22, F.P.O. New York 09593, Daily, 2,600, 4 pages, 11 x 17.

Hawaii Marine, Joint Public Affairs Office, Kaneohe Marine Corps Air Station, F.P.O. San Francisco 96615, Weekly, 8,000, 8 pages, 10 x 15.

Independence Guardian, U.S.S. Independence (CV-62), F.P.O. New York 09501, Daily (at sea), 1,000, 8 pages, 8½ x 11.

*The Jack Tar, Box 27, U.S. Naval Station, Rota, Spain, F.P.O. New York 09540, Weekly, 6,000, 8 pages, 11½ x 15, $.69/line.

KA LEO O KILAUEA, USS Kilavea AE-26, F.P.O. San Francisco 96601, Biweekly, 340, 4 pages, 8½ x 11.

Kangaroo Kourier, Public Affairs Office, NMCB-133, F.P.O. New York 09501, 750, 20 pages, 10½ x 8.

Ledger, USS Enterprise CVW-65, F.P.O. San Francisco 96601, Daily (at sea), 1,200, 8 pages, 8 x 10.

Mira Que Pasa, Box 1, F.P.O. New York 09551, Biweekly, 1,800, 10 pages, 17 x 22.

The Mountain Signal, U.S.S. Mount Whitney Lee 20, F.P.O. New York 09501, 800, 12 pages.

*Panorama, Box 40, NSA, F.P.O. New York 09521, Weekly, 5,000-8,000, 8 pages, 5 col., $.45/line.

Puget Soundings, U.S.S. Puget Sound (AD-38), F.P.O. New York 09501, Quarterly, 2,000, 24 pages, 8 x 10.

Pulse, Public Affairs Office, USS Forrestal (CV-59), F.P.O. New York 09501, Monthly, 5,900, 20 pages, 8 x 10.

PWC Journal, U.S. Navy Public Works Center, F.P.O. San Francisco 96630, Monthly, 1,300, 16 pages, 8 x 10½.

The SASEBO News, Public Affairs Office, Box 271, US NAVORDFAC, F.P.O. Seattle 98766, Weekly, 400, 10 pages, 8 x 13.

Seahawk, CFAY Special Services, F.P.O. Seattle 98762, Weekly, 8,000, 8 pages, 5 col.

The Talking Stick, NAVCOMMSTA Harold E. Holt, Box 8, F.P.O. San Francisco 96680, Weekly, 600, 8 pages, 10 x 15.

Tartan Log, PAO, NAVSECGRUACT, F.P.O. New York 09518, Monthly, 700, 8 x 10½.

Torii Teller, Public Affairs Office, MCAS, F.P.O. Seattle 98764, Weekly, 3,000, 12 pages, 11¼ x 8¼.

White Falcon, U.S. Naval Station, F.P.O. New York 09571, 4,000, 6 pages, Tabloid.

The Windjammer, Public Affairs Office, CFAO/NAFK, F.P.O. Seattle 98770, Monthly, 1,500, 16 pages, 8 x 10.

Europe

*Praetorian, Stampa Generale S.R.L., Via S. Andrea della Fratte 36, 00187, Rome, Italy, Weekly, 5,000, Tabloid, $.28/line, Adv. to Stampa Generale S.R.L., Via S. Andrea della Fratte 36, 00187, Rome, Italy.

*Armed Forces Network TV-Guide, Cmb. H, Darmstaedter Landster 199, 600 Frankfurt/Main 70, West Germany, 92,000, Monthly, 5½ x 7⅝, $1,880/page, Adv. to N. DeFillippes Co., 420 Lexington Ave., New York, NY 10017, (212) 697-8185.

*R & R Entertainment Digest with Guide to TV, Werbe Cmb H, Engelstrasse 6, D-6900 Heidelberg, West Germany, Monthly, 84,300, 5¾ x8¼, $1,600/page.

*Overseas Life, UTTO-Hahn-Strasse 9, 6382 Friedrichsdorf 2, Koppern, West Germany, 80,890, Monthly, 8¼ x 11, $2,260/page.

***Publications marked with an asterisk accept advertising. Among the newspapers which accept advertising, most are represented by Military Media, Inc., 257 Central Park West, N.Y. 10024.**

A

ABA Newsletter	San Diego, CA
APG News	Aberdeen, MD
Acadian	Winter Harbor, ME
Accent	Ft. Huachuca, AZ
Acorn	Oakland, CA
Adak Eagle's Call	FPO Seattle
Admissions Ofc. Info. Bulletin	West Point, NY
Aegean Breeze	APO New York
Aerospace Historian	Manhattan, KS
Aerospace Safety	Norton AFB, CA
AFCS Frequency Mgmt. Digest	Scott AFB, IL
AFCS Intercom	Scott AFB, IL
AFROTC Education Journal	Maxwell AFB, AL
Air Defense Magazine	Ft. Bliss, TX
Air Force Comptroller	Washington, DC
Air Force Engineering & Svcs. Quarterly	Wright Patterson AFB, OH
Air Force Law Review	Maxwell AFB, AL
Air Force Magazine	Washington, DC
Air Force Officers Career Newsletter	Randolph AFB, TX
Air Force Times	Washington, DC
Air Force University Review	Maxwell AFB, AL
Air Pulse	Omaha, NE
Air Reservist	Washington, DC
Air Scoop	Lakehurst, NJ
Air Scoop	Little Rock, AFB, AR
Air Weather Service Observer	Scott AFB, IL
Aircraftsman	Corpus Christi, TX
Airlifter	Dover AFB, DE
Airlifter	Sarasota, FL
Airman	Washington, DC
Airscoop	Enid, OK
Airwinger	Jacksonville, FL
Alaskan Pioneer	Ft. Richardson, AK
Aloha Hoalauna	Pearl Harbor, HI
All Hands Magazine	Washington, DC
Altus Galaxy	Altus, OK
American Ex-Prisoners of War	Springhill, LA
American Forces Press Svc. Clipsheet	Arlington, VA
American Legion Auxiliary Natl. News	Indianapolis, IN
American Legion Magazine	Indianapolis, IN
American Legion Press Assoc.	Tonawanda, NY
American Veterans Com. Bulletin	Washington, DC

American War Mothers .... Washington, DC
AMVETS Newsletter .... Washington, DC
Anatolian Times .... APO New York
Anchor .... Floral Park, NY
Antenna .... APO New York
Armed Forces Journal International .... Washington, DC
Armed Forces Network TV-Guide .... Frankfurt, West Germany
Armed Services News .... Virginia Beach, VA
Armor .... Ft. Knox, KY
Army .... Washington, DC
Army Administrator Magazine .... Ft. Benjamin Harrison, IN
Army & Navy Union Newsletter .... Lakemore, OH
Army Dollar .... Indianapolis, IN
Army Flier .... Ft. Rucker, AL
Army Host .... Washington, DC
Army Logistician Magazine .... Ft. Lee, VA
Army Museum Newsletter .... Washington, DC
Army-Navy-Air Force Courier .... San Antonio, TX
Army Navy Store .... New York, NY
Army Personnel Letter .... Washington, DC
Army Reserve Magazine .... Washington, DC
Army Times .... Washington, DC
Arradcom Voice .... Dover, NJ
Arrow .... APO New York
Artician .... APO Seattle
Auxiliare .... Kansas City, MO

B

Banner .... Lawrenceville, NJ
Banner .... Philadelphia, PA
Baygram .... San Francisco Bay, CA
Bayonet .... Columbus, GA
Bayonet .... Ft. Ord, CA
Beacon .... Lexington, KY
Beacon .... Riverside, CA
Beam .... Corpus Christi, TX
Beamrider .... Indianapolis, IN
Beehive .... Beeville, TX
Bellringer .... Ft. Baker, CA
Benchmark .... Camp Pendleton, CA
Bergstrom AFB Guide .... Bergstrom AFB, TX
Berlin Observer .... APO New York
Big E Magazine .... FPO San Francisco
Blazer .... APO New York

Blue Falcon Monthly .... Ft. Bragg, NC
Blue Print.. Columbus, MS
Blue Ridger .... Richmond, VA
Bluejacket .... Millington, TN
Blythe Spirit.......................................Blytheville AFB, AR
Bolling Beam ....Bolling AFB, DC
Bomb Blast ....APO New York
Booster ....Williamsburg, VA
Boot ....Parris Island, SC
Bow Hook .... Charleston, SC
Brigadier .... Charleston, SC
Bugle Call... Bridgeport, OH
Bullseye ....North Las Vegas, NV

C

Cactus Comet .... Yuma, AZ
Caducean ....Tripler Army Med. Cntr., HI
Camp Lejeune Globe .... Camp Lejeune, NC
Camp Lejeune Trader............................ Camp Lejeune, NC
Can Do .... Carpentersville, IL
Cannoneer .... Ft. Sill, OK
Capital Flyer .... Andrews AFB, MD
Careergram .... Washington, DC
Carry On .... Washington, DC
Carswell Sentinel................................ Carswell AFB, TX
Casemate .... Ft. Hamilton, B'klyn, NY
Castle .... Annapolis, MD
Catholic War Veterans/USA................... Washington, DC
Centergram .... Pearl Harbor, HI
Centerline... Bethesda, MD
Centerline.. San Diego, CA
Champlaner .... Plattsburgh, NY
Channel .... Oceanside, CA
Chanute This Week .... Rantoul, IL
Chaplain Corps Items of Interest .... Washington, DC
Charleston Airlift Dispatch .... Charleston AFB, SC
Cherry Point Windsock .... Camp Lejeune, NC
Chevron .... San Diego, CA
Chief of Supply Newsletter .... Hickam AFB, HI
Civilian Courier .... Tyndall AFB, FL
Civilian Guidepost .... Camp Lejeune, NC
Civilian News Briefs .... Athens, GA
Civilian Personnel Newsletter .... Luke AFB, AZ
CIVINFO.......................................Twentynine Palms, CA

Classmate.. Monterey, CA
Club & Food Svc. .... Lynbrook, NY
Club Executive .... Washington, DC
COG "I" Bulletin...................................... Philadelphia, PA
Command Post Newspaper .... O'Fallon, IL
Command Review................................ APO San Francisco
Commissary Newsletter .... Albany, GA
Communications................................ Peterson AFB, CO
Communicator .... Herlong, CA
Community Life .... APO New York
Comnavairpac.......................................San Diego, CA
Comsubpac .... Pearl Harbor, HI
Continental Marine ....New Orleans, LA
Contrails .... Wichita, KS
Cootie Courier.. Fostoria, OH
Crossfeed .... Denver, CO
Crossroads .... APO New York
Cutler Sparks..................................... East Machias, ME

D

Dad Talk .... Topeka, KS
Daedalus Flyer .... San Antonio, TX
DATC FMAG "Dialog" ....San Diego, CA
Dateline .... Ft. Hood, TX
DCASR Dallas Quality Assurance Newsletter .... Dallas, TX
DCASR News .... Chicago, IL
Daughters of Union Veterans of Civil War/Genl. Orders ... Springfield, IL
Daughters of Union Veterans/Historical Reporter......... Springfield, IL
Deckplate.. Washington, DC
Defender .... Hancock Field, NY
Defender .... APO San Francisco
Defense Mgmt. Journal .... Alexandria, VA
Defense Transportation Journal .... Washington, DC
Depot Dispatch.................................. Trumansburg, NY
DESC Customer Information Bulletin.................. Dayton, OH
Desert Airman .... Tucson, AZ
Desert RoundUp....................................... Fallon, NV
Desert Sampler .... Dugway, UT
Desert Trail .... Twentynine Palms, CA
DFSC Fuel Line .... Alexandria, VA
Desert Wings .... California City, CA
Digest .... Long Beach, CA
Digest .... New Orleans, LA
Direction Magazine .... Washington, DC

Disabled American Veterans/Auxiliary Magazine .... Cold Spring, KY
Disabled American Veterans Magazine .... Cold Spring, KY
Disabled Officers Assoc. .... Washington, DC
Discovery .... Brooks AFB, TX
Dispatch .... Maxwell AFB, AL
Distaff Side .... Coronado, CA
Dockside Gosport .... Portsmouth, VA
Dolphin .... Groton, CT
Donau Republic .... APO New York
Dragoon .... APO New York
Driver .... Norton AFB, CA
Dry Dock .... San Diego, CA
Dyess Peacemaker .... Dyess AFB, TX

E

Eagle and Swan .... New York, NY
Eastern Arrow .... Bayonne, NJ
EEO Newsletter .... Washington, DC
Eglin Eagle .... Ft. Walton Beach, FL
Eielson AFB Goldpanner .... Eielson AFB, AK
1812 Ancestors Index & Roster .... Washington, DC
El Tiburon .... FPO New York
Emblem .... Albany, GA
Emphasis and Trends .... Norfolk, VA
Encampment .... Lakemore, OH
Engineer .... Ft. Belvoir, VA
Eur Army Magazine .... APO New York
European Scientific Notes .... FPO New York
Examiner .... Ft. Gordon, GA
Exchange & Commissary News .... Lynbrook, NY
Extension Course Institute Catalog .... Gunter AFB, AZ

F

Faceplate .... Washington, DC
Fairchild Times .... Fairchild AFB, WA
Falconews .... USAF Academy, CO
Family .... New York, NY
Family Housing Newsletter .... Lemoore, CA
Family Svcs. Mirror .... Offutt AFB, NE
Far East Network Newsletter .... APO San Francisco
Field Artillery Journal .... Ft. Sill, OK
Fifteenth Air Base Wing/Personnel Newsletter .... Hickam AFB, HI
Fighter Forum .... Homestead, FL

Five x Five .... APO New York
Flight Jacket .... Laguna Hills, CA
Flyer .... Langley AFB, VA
Focus .... Albuquerque, NM
Focus on Four .... FPO San Francisco
Fog Horn .... San Francisco, CA
Food News .... Philadelphia, PA
Ft. Bliss Monitor .... El Paso, TX
Ft. Campbell Courier .... Ft. Campbell, KY
Ft. Devens Dispatch .... Ft. Devens, MA
Ft. Dix Post .... Mount Holly, NJ
Ft. Gordon Rambler .... Augusta, GA
Ft. Hood Sentinel .... Temple, TX
Ft. Jackson Leader .... Ft. Jackson, SC
Ft. Leavenworth Lamp .... Ft. Leavenworth, KS
Ft. Ord Panorama .... Pacific Grove, CA
Ft. Riley Post .... Junction City, KS
Ft. Sam Houston News Leader .... Universal City, TX
Ft. Sheridan Tower .... Grayslake, IL
Fortitudine .... Washington, DC
Forty & Eighter .... Indianapolis, IN
14th Engineer Battalion Dimension .... Ft. Ord, CA
Frankfurt Chronicle .... APO New York
Fridaygram .... Washington, DC
Frontline .... APO New York
Fuji Flyer .... APO San Francisco
Full Deflection .... Houston, TX

G

Gateway Reporter .... St. Louis, MO
Gator .... Norfolk, VA
Gauntlet .... APO San Francisco
Gemini .... Malmstrom AFB, MT
General Society of Colonial Wars Gazette .... Glendale, OH
Giron .... Miami, FL
Globetrotter .... Redlands, CA
Golden Eagle .... Lemoore, CA
Golden Griffon .... Charlotte, NC
Golden Sphinx .... Dayton, OH
Gold Star Mothers .... Washington, DC
Goodfellow Monitor .... San Angelo, TX
Gosport .... Pensacola, FL
Great Lakes Bulletin .... Grayslake, IL
Grissom Contact .... Greentown, IN
Guantanamo Gazette .... FPO New York

H

Hansconian .... Hanscom AFB, ME
Harrison Post .... Ft. Benjamin Harrison, IN
Hawaii Marine .... FPO San Francisco
Hawaii Navy News .... Pearl Harbor, HI
Hawaiian Falcon .... Hickam AFB, HI
Hercules Herald .... Pope AFB, NC
Hi-Desert Flyer .... George AFB, CA
High Flyer .... Ft. Bliss, TX
Hill Top Times .... Hill AFB, UT
Hoist .... San Diego, CA
Hotline .... Washington, DC
Howling Gale .... New London, CT
Huachuca Scout .... Ft. Huachuca, AZ
HUB .... Roy, UT

I

Independence Guardian .... FPO New York
Infantry Magazine .... Ft. Benning, GA
Inside the Turret .... Ft. Knox, KY
Inspector General Brief .... Washington, DC
Interceptor Magazine .... Peterson AFB, CO
Intercom .... Richards-Gebaur AFB, MO
Intercom .... Sacramento, CA
Intercom .... Willow Grove, PA
International Defense Review .... Hollywood, CA
Ironsides .... APO, New York

J

Jack Tar .... FPO New York
JAG Journal .... Washington, DC
JAG Reporter .... Washington, DC
JAX Air News .... Jacksonville, FL
Jet Gazette .... Bergstrom AFB, TX
Jet Gazette .... Williams AFB, AZ
Jet Observer .... Virginia Beach, VA
Jet Scope .... Tyndale AFB, FL
Jetster .... McConnell AFB, KS
Jewish Veterans .... Washington, DC
Jewish War Veterans/USA Bulletin .... Washington, DC
Jing Bao Journal .... Selden, NY
Journal of the Army Intelligence & Security Command..... Arlington, VA

K

Ka Leo O Kilauea .... FPO San Francisco
Kadena Falcon .... APO San Francisco
Kangaroo Kourier .... FPO New York
Keesler News .... Keesler AFB, MS
Kelly Observer .... San Antonio, TX
Ketch 22 .... Seattle, WA
Keynotes .... Keyport, WA
Kingsley Times .... Kingsley Field, OR
Kisatchie Guardian .... Ft. Polk, LA
Knight Express .... Adel, GA

L

Lackland Tailspinner .... Lackland AFB, TX
Ladycom .... Washington, DC
Ladies of the Grand Army of the Republic/Journal .... Bridgeport, OH
Langley Outlook .... Langley AFB, VA
Leader .... Finley, ND
Leatherneck Magazine .... Quantico, VA
Ledger .... FPO San Francisco
Legion of the Guardsmen Auxiliaries .... Parkersburg, WV
Legion of Guardsmen/General Orders .... Parkersburg, WV
Legions of Guardsmen Magazine .... Parkersburg, WV
Legion of Valor of USA .... Arlington, VA
Lemot .... Pensacola, FL
Limelite .... Loring AFB, ME
Log .... Annapolis, MD
Long Beach Dispatch .... San Diego, CA
Lowry Airman .... Aurora, Co

M

MAC Flyer .... Scott AFB, IL
Mach Meter .... Cannon AFB, NM
Main Megaphone .... APO New York
Maintainer .... APO San Francisco
Maintenance Magazine .... Norton AFB, CA
Manpower & Organization Newsletter .... Randolph AFB, TX
Mare's Tale .... Vallejo, CA
Marine Corps Club System Newsletter .... Washington, DC
Marine Corps Gazette .... Quantico, VA
Marine's Magazine .... Arlington, VA
Marmara Messenger .... APO New York

Masthead .... San Francisco, CA
Mather AFB Wing Tips .... Rancho Cordova, CA
McClellan AFB Spacemaker .... Rancho Cordova, CA
McClellan News .... Ft. McClellan, AL
MCLSBLANT Newsletter .... Albany, GA
McGuire Airtides .... Pemberton, NJ
Mechanews .... Mechanicsburg, PA
Medcom Examiner .... APO New York
Medical Patriot .... San Antonio, TX
Mesa Missilier .... Lompoc, CA
Midwest Reporter .... Shawnee Mission, KS
Military Affairs .... Manhattan, KS
Military Airlift Command Info. Newsletter .... Scott AFB, IL
Military Chaplain .... Washington, DC
Military Engineer .... Washington, DC
Military Intelligence Magazine .... Ft. Huachuca, AZ
Military Journal .... Bennington, VT
Military Market Magazine .... Washington, DC
Military Media Review .... Ft. Benjamin Harrison
Military Medicine .... Kensington, MD
Military Order of the Cootie/Bulletin .... St. Joseph, MO
Military Order of the Cootie/General Orders .... St. Joseph, MO
Military Order of Foreign Wars of USA/Newsletter ..... Philadelphia, PA
Military Order of Purple Heart Newsletter/Ladies Auxiliary Baltimore, MD
Military Police Law Enforcement Journal .... Ft. McClellan, AL
Military Railway Svc. Veterans .... Pass Christian, MS
Military Review .... Ft. Leavenworth, KS
Military Sun Press .... Honolulu, HI
Mini-Gram .... Adelphi, MD
Minuteman .... Malmstrom AFB, MT
Mira Que Pasa .... FPO New York
Mirror .... Mayport, FL
Misawan .... APO San Francisco
Missile .... Pt. Mugu, CA
Missileer .... Patrick AFB, FL
Moffett News .... Sunnyvale, CA
Mohawk Flyer .... Holland Patent, NY
Money Matters .... Charleston, SC
Monitor .... Whiteman AFB, MO
Monmouth Message .... Ft. Monmouth, NJ
Monthly Memo .... Virginia Beach, VA
Motor Mouth .... Detroit, MI
Mountain Signal .... FPO New York
Mountaineer .... Ft. Carson, CO
Mustin Missile .... Pensacola, FL

N

NADC Reflector .... Warminster, PA
NAFI Notes .... Tyndall AFB, FL
Narfacts .... San Diego, CA
Nartopix .... Washington, DC
National AMVET .... Washington, DC
National Assoc. of Concerned Veterans .... Washington, DC
National Defense Magazine .... Washington, DC
National Order of Trench Rats .... Downey, CA
National Organization Bulletin .... Kansas City, MO
National PLAV News .... Chicago, IL
National Woman's Relief Corps/General Orders .... Springfield, IL
Navagator .... Orlando, FL
Naval Affairs .... Washington, DC
Naval Affairs Magazine .... Washington, DC
Naval Air Reserve Notes .... Pt. Mugu, CA
Naval Aviation News .... Washington, DC
Naval Civil Engineer .... Pt. Hueneme, CA
Naval Engineers Journal .... Washington, DC
Naval General Library Svcs. Newsletter .... Pensacola, FL
Naval Oceanographic Ofc. Bulletin .... Bay St. Louis, MS
Naval Research Reviews .... Arlington, VA
Naval Reservist News .... New Orleans, VA
Naval Station Newsletter .... Annapolis, MD
Naval War College Review .... Newport, RI
Navelex News .... Portsmouth, VA
Navigator Magazine .... Mather AFB, CA
Navmiro .... Philadelphia, PA
NAVSEACENLANT .... Portsmouth, VA
Navy Campus Newsletter .... Pensacola, FL
Navy Editor Svc. .... Arlington, VA
Navy Family Housing News .... Oakland, CA
Navy Foodservice .... Washington, DC
Navy Human Resource Journal .... Millington, TN
Navy Lifeline .... Norfolk, VA
Navy Mothers' News .... Phoenix, AZ
Navy News .... Arlington, VA
Navy News .... San Diego, CA
Navy Policy Briefs .... Washington, DC
Navy Resale System News Digest Magazine .... Brooklyn, NY
Navy Supply Corps Newsletter .... Washington, DC
Navy Times .... Washington, DC
Navy Wives News .... Virginia Beach, VA
NCOA Journal .... San Antonio, TX

Newport Navalog .... Newport, RI
Newsbeam .... APO New York
Newsgram .... Charleston, SC
Newsletter for Retired Marines .... Washington, DC
Nolair News .... New Orleans, LA
North Islander .... Encinitas, CA
Northern Light .... APO San Francisco
Northern Light .... Marquette, MI
Northern Scanner .... APO New York
Northern Airlifter .... Tacoma, WA
Note Book .... New York, NY
NSA Newsletter .... Brooklyn, NY

O

Oak Leaf .... Silver Springs, MD
Observation Post .... Twentynine Palms, CA
Observer .... Barksdale AFB, LA
Observer .... Norad, CO
Off Duty .... Newport Beach, CA
Officer .... Washington, DC
Officer Personnel Newsletter .... Washington, DC
Officers Wives League Bulletin .... Pensacola, FL
Offutt & SAC Base Guide .... Omaha, NE
On Target .... Heath, OH
Oracle .... APO New York
Order Wire .... Ft. Ritchie, MD
Outlook .... Patuxant River, MD
Outlook .... San Diego, CA
Overseas Life .... Koppern, West Germany

P

Pacific Health Bulletin .... Pearl Harbor, HI
Panorama .... FPO New York
Parameters .... Carlisle Barracks, PA
Paraplegia News .... Washington, DC
Patriot .... Ft. Stewart, GA
Patroller .... Brunswick, ME
Pearl Harbor Gram .... Portland, OR
Pendleton Scout .... Camp Pendleton, CA
Pentagram News Periscope .... San Diego, CA
Pershing Cable .... APO New York
Personnel Memo .... Los Angeles, CA
Personnel Newsletter .... Robins AFB, GA

Persuader .... Ft. Bragg, NC
Phantom Forum .... APO New York
Photogram .... APO New York
Photoscope .... Panama City, FL
Pillars and Post .... APO New York
Pirmasens Com. Times .... APO New York
Plainsman .... Ellsworth AFB, SD
Pointer .... Jacksonville, NC
Pointer Magazine .... West Point, NY
Polaris .... Elmhurst, IL
Port Reporter .... APO New York
Praetorian .... Rome, Italy
Procurement Newsletter .... Washington, DC
Profile .... Indiana Head, MD
Prospector .... Barstow, CA
PS Magazine .... Lexington, KY
PT Boat All Hands .... Memphis, TN
Puget Soundings .... Bremerton, WA
Puget Soundings .... FPO New York
Pulaski County Democrat .... Waynesville, MO
Pulse .... FPO New York
Pupukahi .... Honolulu, HI
Purple Heart .... Arlington, VA
Purple Heart Magazine .... Baltimore, MD
PWC Journal .... FPO San Francisco

Q

Quan .... Hurley, NY
Quantico Sentry .... Dunfries, VA

R

R & R Entertainment Digest .... Heidelberg, West Germany
Raider Magazine .... APO New York
Ranger .... Ft. Lewis, WA
Ranger Journal .... Seabrook, TX
Ready Letter .... Great Lakes, IL
Recon .... Philadelphia, PA
Recreational Svcs. Today .... Washington, DC
Recruiter .... Arlington, VA
Red River Review .... Texarkana, TX
Redstone Rocket .... Huntsville, AL
Regional Reflections .... Philadelphia, PA
Relay .... APO San Francisco

Reliance .... Washington, DC
Report of NRL Progress .... Washington, DC
Reserve Memphis News .... Memphis, TN
Retired Naval Personnel Newsletter .... Washington, DC
Retired Officer Magazine .... Washington, DC
Review .... Washington, DC
Retreads Blowout .... Flushing, NY
Retreads/Roster .... Flushing, NY
Rheinlander .... APO New York
Richards Gebaur Communicator .... Richards Gebaur AFB, MO
Rickenbacker AFB Skyhawk .... Rickenbacker AFB, OH
Ritchie News .... Ft. Ritchie, MD
Robins Rev-Up .... Robins AFB, GA
Rocket .... Hawthorne, NV
Rocketeer .... China Lake, CA
Rotovue .... Jacksonville, NC
Roundup .... Lubbock, TX

S

SAC Personnel Brief .... Offutt AFB, NE
SAM .... Arlington, VA
Safety Sentinel .... New Orleans, LA
St. Louis Quality Assur. Newsletter .... St. Louis, MO
Salutations .... Indianapolis, IN
Salvo .... Watervliet, NY
San Diego Dispatch .... San Diego, CA
Sandpaper .... El Centro, CA
Sar-Tar News .... Norfolk, VA
SASEBO News .... FPO Seattle
Sascom Courier .... APO New York
Sea Power .... Upper Marlboro, MD
Sea Power .... Washington, DC
Seabee Courier .... Gulfport, MS
Seabee Coverall Newspaper .... Pt. Hueneme, CA
Seacoast Flyer .... Kennebunk, ME
Seacoast Flyer .... Pease AFB, NH
Seahawk .... FPO Seattle
Sentinel .... Jonesboro, GA
Sentinel .... Warrn AFB, WY
Sentry .... Minot AFB, ND
Svc. Star Newsletter .... Shandyside, OH
7th World .... Ft. Eustis, VA
Scope .... Seymour Johnson AFB, NC
Sharpshooter .... Tracy, CA

Shaw AFB Recon .... Cayce, SC
Sheppard Senator .... Sheppard AFB, TX
Shipmate .... Annapolis, MD
Shock & Vibration Digest .... Washington, DC
Shore Leaf .... San Diego, CA
Shore Leaf .... Encinitas, CA
Sky Ranger .... Dallas, TX
Sky Anchor .... Key West, FL
Skyhawk .... Grove City, OH
Skyline .... Meridean, MS
Skywrighter .... Dayton, OH
Smoke Signal .... Mt. Clemens, MI
Soesterberg .... APO New York
Sof Hi-lites .... Hurlburt Field, FL
Sojourner .... Alexandria, VA
Soldiers .... Alexandria, VA
Sons of Confederate Veterans .... Hattiesburg, MS
Soundoff .... Annapolis, MD
Sound Off .... Milwaukee, WI
Soundings .... Norfolk, VA
Sourdough Sentinel .... Elmendorf AFB, AK
Southerner .... Charleston, SC
Sou'wester .... San Diego, CA
Space Sentinel .... Yuba City, CA
Spangled Banner .... APO New York
Spearhead .... APO New York
Standard Bearer .... Garden City, NY
Star .... APO New York
Star Presidian .... San Francisco, CA
Strand Sentry .... Myrtle Beach AFB, SC
Stars & Stripes .... APO New York
Stripe .... Washington, DC
Sunburst .... Alamogordo, NM
Sunburst .... Ft. Hood, TX
Sun Ship .... Pascagoula, MS
Supplement to the AF Policy for Commanders .... Washington, DC
Supply Chest .... Norfolk, VA
Surveyor .... Washington, DC
Systems Management Newsletter .... Andrews AFB, MD

T

TAC Attack .... Langley AFB, VA
Tactical Communicator .... Scott AFB, IL
Tailwind .... Fairfield, CA

Talking Stick .... FPO San Francisco
Talon .... USAF Academy, CO
Tartan Log .... FPO New York
Tester .... Patuxent River, MD
Texas Veterans News .... Ft. Worth, TX
Thunderbolt of MacDill .... Tampa, FL
Tiger Log .... Burbank, CA
Tiger Talk .... Pineville, LA
Tobyhanna Reporter .... Tobyhanna, PA
Torch .... Waterbury, CT
Torch, Nat. Ladies Aux. .... Alexandria, VA
Torch, Vet. of WWI .... Alexandria, VA
Torii Teller .... FPO Seattle
Tortt .... APO San Francisco
Trade Winds .... Honolulu, HI
Translog Magazine .... Washington, DC
Traveler .... APO New York
Triad .... San Diego, CA
Trident .... Charleston, SC
Tropic Lightning News .... Schofield Barracks, HI
Tyndall Guide & Directory .... Panama City, FL
Typhoon Topics .... APO San Francisco

U

UDC Magazine .... Richmond, VA
Underseer .... Panama City, FL
USAF Academy Catalog .... USAF Academy, CO
USAF Interceptor Weapons Newsletter .... Tyndall AFB, FL
USAF Medical Svc. Digest .... Washington, DC
USAF News for Retired Personnel .... Randolph AFB, TX
US Air Force Policy Letter for Commanders .... Washington, DC
US Army Ambulance Svc. Assoc. Bulletin .... Huntingdon Valley, PA
US Army Aviation Digest .... Ft. Rucker, AL
US Army Recruiting and Career Counseling Journal .... Ft. Sheridan, IL
US Naval Institute Proceedings .... Annapolis, MD
US Navy Chaplains Bulletin .... Washington, DC
US Navy Medicine .... Washington, DC
US Navy Security Assistance Newsletter .... Washington, DC
USACSC Interface .... Ft. Belvoir, VA
USAFE Civilian Personnel Newsletter .... APO New York

V

Valley Bomber .... Castle AFB, CA

Vanguard Magazine .... San Antonio, TX
Veterans of WWI/General Orders .... Alexandria, VA
VFW Magazine .... Kansas City, MO
VFW Ladies Auxiliary .... Kansas City, MO
Vigileer .... APO New York
Volunteer .... Alameda, CA

W

War Cry of 1812 .... Washington, DC
Washington Action Reporter .... Kansas City, MO
Water Spectrum .... Washington, DC
West Virginia Legionnaire .... Charleston, WV
Western Arrow .... Oakland Army Base, CA
Westword .... San Bruno, CA
What's Sup .... Washington, DC
Wheel .... Norfolk, VA
Whidbey Crosswind .... Oak Harbor, WA
White Falcon .... FPO New York
White Sands Missile Ranger .... White Sands Missile Range, NM
Whiting Tower .... Milton, FL
Widows of WWI Newsletter .... Washington, DC
Windjammer .... FPO Seattle
Wing Spread .... Mountain Home AFB, ID
Wurtsmith News .... Wurtsmith AFB, MI

Y

Yukon Sentinel .... Ft. Wainwright, AK

Z

Zoeckler Zephyr .... APO San Francisco